Site Planning & Design ARE Mock Exam

(SPD of Architect Registration Exam)

ARE Overview, Exam Prep Tips,
Multiple-Choice Questions and Graphic Vignettes,
Solutions and Explanations

Gang Chen

ArchiteG®, Inc.
Irvine, California

Site Planning & Design ARE Mock Exam (SPD of Architect Registration Exam): ARE Overview, Exam Prep Tips, Multiple-Choice Questions and Graphic Vignettes, Solutions and Explanations

V1.10 Incorporated minor revisions on 1/1/2015

Copy Editor: Terri McClain

ArchiteG®, Inc.
http://www.ArchiteG.com

ISBN: 978-1-61265-011-1

PRINTED IN THE UNITED STATES OF AMERICA

Dedication

To my parents, Zhuixian and Yugen,
my wife, Xiaojie, and my daughters,
Alice, Angela, Amy, and Athena.

Disclaimer

Site Planning & Design ARE Mock Exam (SPD of Architect Registration Exam) provides general information about Architect Registration Exam. The book is sold with the understanding that neither the publisher nor the authors are providing legal, accounting, or other professional services. If legal, accounting, or other professional services are required, seek the assistance of a competent professional firm.

The purpose of this publication is not to reprint the content of all other available texts on the subject. You are urged to read other materials and tailor them to fit your needs.

Great effort has been taken to make this resource as complete and accurate as possible. However, nobody is perfect and there may be typographical errors or other mistakes present. You should use this book as a general guide and not as the ultimate source on this subject. If you find any potential errors, please send an e-mail to:
info@ArchiteG.com

Site Planning & Design ARE Mock Exam (SPD of Architect Registration Exam) is intended to provide general, entertaining, informative, educational, and enlightening content. Neither the publisher nor the author shall be liable to anyone or any entity for any loss or damages, or alleged loss or damages, caused directly or indirectly by the content of this book.

If you do not wish to be bound by the above, you may return this book to the publisher for a full refund.

Legal Notice

ARE Mock Exam series by ArchiteG, Inc.

Time and effort is the most valuable asset of a candidate. How to cherish and effectively use your limited time and effort is the key of passing any exam. That is why we publish the ARE Mock Exam series to help you to study and pass the ARE exams in the shortest time possible. We have done the hard work so that you can save time and money. We do not want to make you work harder than you have to.

Do not force yourself to memorize many numbers. Read through the numbers a few times, and you should have a very good impression of them.

You need to make the judgment call: If you miss a few numbers, you can still pass the exam, but if you spend too much time drilling these numbers, you may miss the big pictures and fail the exam.

The official NCARB sample MC questions have no explanations, and the official NCARB sample vignettes have no step-by-step solutions. The existing ARE practice questions or exams by others are either way too easy or way over-killed. They do NOT match the real ARE exams at all.

We have done very comprehensive research on the official NCARB guides, many related websites, reference materials, and other available ARE exam prep materials. We match our mock exams vignettes questions as close as possible to the NCARB samples and the real ARE exams instead. Some readers had failed an ARE exam two or three times before, and they eventually passed the exam with our help.

All our books include a step-by-step solution to each of the NCARB sample vignette with screenshots and related NCARB software commands, a complete set of MC questions (except for the SD ARE mock exam) and vignettes matching the real ARE exams, including number of questions, format, type of questions, etc. We also include detailed answers and explanations to our MC questions, and a step-by-step solution to each of the mock vignettes with screenshots and related NCARB software commands. Our DWG files for our mock exam vignettes can also be installed and used with the NCARB software.

There is some extra information on ARE overviews and exam-taking tips in Chapter One. This is based on NCARB AND other valuable sources. This is a bonus feature we included in each book because we want our readers to be able to buy our ARE mock exam books together or individually. We want you to find all necessary ARE exam information and resources at one place and through our books.

All our books are available at
http://www.GreenExamEducation.com

How to Use This Book

We suggest you read *Site Planning & Design ARE Mock Exam (SPD of Architect Registration Exam)* at least three times:

Read once, and cover Chapter One, Two and Appendixes and the related FREE PDF files and other resources. Highlight the information with which you are not familiar.

Read a second time, this time focusing on the highlighted information to memorize. You can repeat this process as many times as you want until you have mastered the content of the book. Pay special attention to the materials listed in Chapter Two, Section B, **the most important documents/publications for SPD division of the ARE exam.**

After reviewing these materials, you can do the mock exam, and then check your answers against the answers and explanations in the back, including explanations for the questions you answered correctly. You may have answered some questions correctly, but for the wrong reason. Highlight the information you are not familiar with.

Like the real exam, the mock exam includes all three types of questions: Select the correct answer, check all that apply, and fill in the blank.

Review your highlighted information, and do the mock exam again. Try to answer 100% of the questions correctly this time. Repeat the process until you can answer all of the questions correctly.

Do the mock exam about two weeks before the real exam, but at least 3 days before the real exam. You should NOT wait until the night before the real exam to do the mock exam: If you do not do well, you will go into panic mode and you will NOT have enough time to review your weakness.

Read for the third time the night before the real exam. Review ONLY the information you highlighted, especially the questions you did not answer correctly when you did the mock exam for the first time.

One important tip for passing the graphic vignette section of the ARE SPD division is to become VERY familiar with the commands of the NCARB software. Many people fail the exam simply because they are NOT familiar with the NCARB software and cannot finish the graphic vignette section within the exam's time limit.

For the graphic vignettes, we include step-by-step solutions, using NCARB Practice Program software, with many screenshots so that you can use this book to become familiar with the commands of the NCARB software, even when you do NOT have a computer in front of you. This book is very light and you can carry it around easily. These two features will allow you to review the graphic vignette section whenever you have a few minutes.

All commands are described in an **abbreviated manner**. For example, **Sketch > Line** means go to the menu on the left hand side of your computer screen, click **Sketch,** and then click **Line** to draw a sketch line. This is typical for ALL commands throughout the book.

The Table of Contents is very detailed, so you can locate information quickly. If you are on a tight schedule, you can forgo reading the book linearly and jump to the sections you need.

All our books including "ARE Mock Exams Series" and "LEED Exam Guides Series" are available at
GreenExamEducation.com

Check out FREE tips and info at **GeeForum.com**, you can post your questions or vignettes for other users' review and responses.

Table of Contents

Chapter One

Overview of the Architect Registration Exam (ARE)

A. First Things First: Go to the website of your architect registration board and read all the requirements for obtaining an architect license in your jurisdiction.
See the following link:
http://www.ncarb.org/Getting-an-Initial-License/Registration-Board-Requirements.aspx

B. Download and Review the Latest ARE Documents at the NCARB Website

1. Important links to FREE and official NCARB documents
The current version of the Architect Registration Exam includes seven divisions:

- Programming, Planning & Practice
- Site Planning & Design
- Building Design & Construction Systems
- Schematic Design
- Structural Systems
- Building Systems
- Construction Documents and Services

Note: *Starting July 2010, the 2007 AIA Documents apply to all ARE Exams.*

Six ARE divisions have a multiple-choice section and a graphic vignette section. The Schematic Design division has NO multiple-choice section, but two graphic vignette sections.

For the vignette section, you need to complete the following graphic vignette(s) based on the ARE division you are taking:

Programming, Planning & Practice
Site Zoning

Site Planning & Design
Site Grading
Site Design

Building Design & Construction Systems
Accessibility/Ramp
Stair Design
Roof Plan

Schematic Design
Interior Layout
Building Layout

Structural Systems
Structural Layout

Building Systems
Mechanical & Electrical Plan

Construction Documents & Services
Building Section

There is a tremendous amount of valuable information covering every step of becoming an architect available free of charge at the NCARB website:
http://www.ncarb.org/

For example, you can find the education guide regarding professional architectural degree programs accredited by the National Architectural Accrediting Board (NAAB), NCARB's Intern Development Program (IDP) guides, initial license, certification and reciprocity, continuing education, etc. These documents explain how you can qualify to take the Architect Registration Exam.

I find the official ARE Guidelines, exam guide, and practice program for each of the ARE divisions extremely valuable. See the following link:
http://www.ncarb.org/ARE/Preparing-for-the-ARE.aspx

You should definitely start by studying the official exam guide and practice program for the ARE division you are taking.

2. **A detailed list and brief description of the FREE PDF files that can be downloaded from NCARB**
   The following is a detailed list of the FREE PDF files that you can download from NCARB. They are listed in order based on their importance.

   - **ARE Guidelines** includes extremely valuable information on the ARE overview, six steps to complete ARE, multiple-choice section, graphic vignette section, exam format, scheduling, sample exam computer screens, links to other FREE NCARB PDF files, practice software for graphic vignettes, etc. You need to read this at least twice.

   - **NCARB Education Guidelines** (Skimming through this should be adequate)

- **Intern Development Program Guidelines** contains important information on IDP overview, IDP steps, IDP reporting, IDP basics, work settings, training requirements, supplementary education (core), supplementary education (elective), core competences, next steps, and appendices. Most of NCARB's 54-member boards have adopted the IDP as a prerequisite for initial architect licensure. This is why you should be familiar with it. IDP costs $350 for the first three years, and then $75 annually. The fees are subject to change, and you need to check the NCARB website for the latest information. Your IDP experience should be reported to NCARB at least every six months and logged within two months of completing each reporting period (the **Six-Month Rule**). You need to read this document at least twice. It has a lot of valuable information.

- **The IDP Supervisor Guidelines** (Skimming through this should be adequate. You should also forward a copy of this PDF file to your IDP supervisor.)

- **Handbook for Interns and Architects** (Skimming through this should be adequate.)

- **Official exam guide, references index, and practice program (NCARB software) for each ARE division**
  This includes specific information for each ARE division. (Just focus on the documents related to the ARE divisions you are currently taking and read them at least twice. Make sure you install the practice program and become very familiar with it. The real exam is VERY similar to the practice program.)

  a. **Programming, Planning & Practice (PPP)**: Official exam guide and practice program for the SPD division
  b. **Site Planning & Design (SPD)**: Official exam guide and practice program (computer software) for the SPD division
  c. **Building Design & Construction Systems (BDCS)**: Official exam guide and practice program for the BDCS division
  d. **Schematic Design (SD)**: Official exam guide and practice program for the SD division
  e. **Structural Systems (SS)**: Official exam guide, references index, and practice program for the SS division
  f. **Building Systems (BS)**: Official exam guide and practice program for the BS division
  g. **Construction Documents and Services (CDS)**: Official exam guide and practice program for the CDS division

- **The Burning Question: Why Do We Need ARE Anyway?** (Skimming through this should be adequate.)

- **Defining Your Moral Compass** (Skimming through this should be adequate.)

- **Rules of Conduct** is available as a FREE PDF file at:

http://www.ncarb.org/
(Skimming through this should be adequate.)

C. The Intern Development Program (IDP)

1. What is IDP?

IDP is a comprehensive training program jointly developed by the National Council of Architectural Registration Boards (NCARB) and the American Institute of Architects (AIA) to ensure that interns obtain the necessary skills and knowledge to practice architecture independently.

2. Who qualifies as an intern?

Per NCARB, if an individual meets one of the following criteria, s/he qualifies as an intern:
a. Graduates from NAAB-accredited programs
b. Architecture students who acquire acceptable training prior to graduation
c. Other qualified individuals identified by a registration board

D. Overview of the Architect Registration Exam (ARE)

1. How to qualify for the ARE

A candidate needs to qualify for the ARE via one of NCARB's member registration boards, or one of the Canadian provincial architectural associations.

Check with your Board of Architecture for specific requirements.

For example, in California, a candidate must provide verification of a minimum of five years of education and/or architectural work experience to qualify for the ARE.

Candidates can satisfy the five-year requirement in a variety of ways:

- Provide verification of a professional degree in architecture through a program that is accredited by NAAB or CACB.

  OR
- Provide verification of at least five years of educational equivalents.

  OR
- Provide proof of work experience under the direct supervision of a licensed architect

2. **How to qualify for an architect license**
   Again, each jurisdiction has its own requirements. An individual typically needs a combination of about eight years of education and experience, as wells as passing scores on the ARE exams. See the following link:
   http://www.ncarb.org/Reg-Board-Requirements

   For example, the requirements to become a licensed architect in California are:
   - Eight years of post-secondary education and/or work experience as evaluated by the Board (including at least one year of work experience under the direct supervision of an architect licensed in a U.S. jurisdiction or two years of work experience under the direct supervision of an architect registered in a Canadian province)
   - Completion of the Comprehensive Intern Development Program (CIDP) and the Intern Development Program (IDP)
   - Successful completion of the Architect Registration Examination (ARE)
   - Successful completion of the California Supplemental Examination (CSE)

   California does NOT require an accredited degree in architecture for examination and licensure. However, many other states do.

3. **What is the purpose of the ARE?**
   The purpose of ARE is NOT to test a candidate's competency on every aspect of architectural practice. Its purpose is to test a candidate's competency on providing professional services to protect the health, safety, and welfare of the public. It tests candidates on the fundamental knowledge of pre-design, site design, building design, building systems, and construction documents and services.

   The ARE tests a candidate's competency as a "specialist" on architectural subjects. It also tests her abilities as a "generalist" to coordinate other consultants' works.

   You can download the exam content and references for each of the ARE divisions at the following link:
   http://www.ncarb.org/are/40/StudyAids.html

4. **What is NCARB's rolling clock?**
   a. Starting on January 1, 2006, a candidate MUST pass ALL ARE sections within five years. A passing score for an ARE division is only valid for five years, and a candidate has to retake this division if she has NOT passed all divisions within the five year period.

   b. Starting on January 1, 2011, a candidate who is authorized to take ARE exams MUST take at least one division of the ARE exams within five years of the authorization. Otherwise, the candidate MUST apply for the authorization to take ARE exams from an NCARB member board again.

   These rules were created by the **NCARB's rolling clock** resolution and passed by NCARB council during the 2004 NCARB Annual Meeting.

5. **How to register for an ARE exam**
   Go to the following website and register:
   http://www.prometric.com/NCARB/default.htm

6. **How early do I need to arrive at the test center?**
   Be at the test center at least 30 minutes BEFORE your scheduled test time, OR you may lose your exam fee.

7. **Exam format & time**
   All ARE divisions are administered and graded by computer. Their detailed exam format and time allowances are as follows:

   1) **Programming, Planning & Practice (PPP)**

Introduction Time:	15 minutes	
MC Testing Time:	**2 hours**	**85 items**
Scheduled Break:	15 minutes	
Introduction Time:	15 minutes	
Graphic Testing Time:	**1 hour**	**Site Zoning (1 vignette)**
Exit Questionnaire:	15 minutes	
Total Time	**4 hours**	

   2) **Site Planning & Design (SPD)**

Introduction Time:	15 minutes	
MC Testing Time:	**1.5 hours**	**65 items**
Scheduled Break:	15 minutes	
Introduction Time:	15 minutes	
2 Graphic Vignettes:	**2 hours**	**Site Grading, Site Design**
Exit Questionnaire:	15 minutes	
Total Time	**4.5 hours**	

   3) **Building Design & Construction Systems (BDCS)**

Introduction Time:	15 minutes	
MC Testing Time:	**1.75 hours**	**85 items**
Scheduled Break:	15 minutes	
Introduction Time:	15 minutes	
3 Graphic Vignettes:	**2.75 hours**	**Accessibility/Ramp, Stair Design, Roof Plan**
Exit Questionnaire:	15 minutes	
Total Time	**5.5 hours**	

4) **Schematic Design (SD)**

Introduction Time:	15 minutes	
Graphic Testing Time:	**1 hour**	**Interior Layout (1 vignette)**
Scheduled Break:	15 minutes	
Introduction Time:	15 minutes	
Graphic Testing Time:	**4 hours**	**Building Layout (1 vignette)**
Exit Questionnaire:	15 minutes	
Total Time	**6 hours**	

5) **Structural Systems (SS)**

Introduction Time:	15 minutes	
MC Testing Time:	**3.5 hours**	**125 items**
Scheduled Break:	15 minutes	
Introduction Time:	15 minutes	
Graphic Testing Time:	**1 hour**	**Structural Layout (1 vignette)**
Exit Questionnaire:	15 minutes	
Total Time	**5.5 hours**	

6) **Building Systems (BS)**

Introduction Time:	15 minutes	
MC Testing Time:	**2 hours**	**95 items**
Scheduled Break:	15 minutes	
Introduction Time:	15 minutes	
Graphic Testing Time:	**1 hour**	**Mechanical & Electrical Plan (1 vignette)**
Exit Questionnaire:	15 minutes	
Total Time	**4 hours**	

7) **Construction Documents and Services (CDS)**

Introduction Time:	15 minutes	
MC Testing Time:	**2 hours**	**100 items**
Scheduled Break:	15 minutes	
Introduction Time:	15 minutes	
Graphic Testing Time:	**1 hour**	**Building Section (1 vignette)**
Exit Questionnaire:	15 minutes	
Total Time	**4 hours**	

8. **How are ARE scores reported?**
All ARE scores are reported as Pass or Fail. ARE scores are processed within 4 to 6 weeks, and sent to your Board of Architecture. Your board then does additional processing and forwards the scores to you.

9. Is there a fixed percentage of candidates who pass the ARE exams?
No, there is NOT a fixed percentage of candidates passing or failing. If you meet the minimum competency required to practice as an architect, you pass. The passing scores are the same for all Boards of Architecture.

10. When can I retake a failed ARE division?
You can only take the same ARE division once within a 6-month period.

11. How much time do I need to prepare for each ARE division?
Every person is different, but on average you need about 40 to 80 hours to prepare for each ARE division. You need to set a realistic study schedule and stick with it. Make sure you allow time for personal and recreational commitments. If you are working full time, my suggestion is that you allow no less than 2 weeks but NOT more than 2 months to prepare for each ARE division. You should NOT drag out the exam prep process too long and risk losing your momentum.

12. Which ARE division should I take first?
This is a matter of personal preference, and you should make the final decision.

Some people like to start with the easier divisions and pass them first. This way, they build more confidence as they study and pass each division.

Other people like to start with the more difficult divisions so that if they fail, they can keep busy studying and taking the other divisions while the clock is ticking. Before they know it, six months has passed and they can reschedule if need be.

Programming, Planning & Practice (PPP) and Building Design & Construction Systems (BDCS) divisions often include some content from the Construction Documents and Service (CDS) division. It may be a good idea to start with CDS and then schedule the exams for SPD and BDCS soon after.

13. ARE exam prep and test-taking tips
You can start with Construction Documents and Services (CDS) and Structural Systems (SS) first because both divisions give a limited scope, and you may want to study building regulations and architectural history (especially famous architects and buildings that set the trends at critical turning points) before you take other divisions.

Complete mock exams and practice questions and vignettes, including those provided by NCARB's practice program and this book, to hone your skills.

Form study groups and learn the exam experience of other ARE candidates. The forum at our website is a helpful resource. See the following link:
http://GreenExamEducation.com/

Take the ARE exams as soon as you become eligible, since you probably still remember portions of what you learned in architectural school, especially structural and architectural history. Do not make excuses for yourself and put off the exams.

The following test-taking tips may help you:

- Pace yourself properly. You should spend about one minute for each Multiple-Choice (MC) question, except for the SS division questions which you can spend about one and a half minutes on.
- Read the questions carefully and pay attention to words like *best, could, not, always, never, seldom, may, false, except,* etc.
- For questions that you are not sure of, eliminate the obvious wrong answer and then make an educated guess. Please note that if you do NOT answer the question, you automatically lose the point. If you guess, you at least have a chance to get it right.
- If you have no idea what the correct answer is and cannot eliminate any obvious wrong answers, then do not waste too much time on the question and just guess. Try to use the same guess answer for all of the questions you have no idea about. For example, if you choose "d" as the guess answer, then you should be consistent and use "d" whenever you have no clue. This way, you are likely have a better chance at guessing more answers correctly.
- Mark the difficult questions, answer them, and come back to review them AFTER you finish all MC questions. If you are still not sure, go with your first choice. Your first choice is often the best choice.
- You really need to spend time practicing to become VERY familiar with NCARB's graphic software and know every command well. This is because the ARE graphic vignette is a timed test, and you do NOT have time to think about how to use the software during the test. If you do not know how, you will NOT be able to finish your solution to the vignette on time.
- The ARE exams test a candidate's competency to provide professional services protecting the health, safety, and welfare of the public. Do NOT waste time on aesthetic or other design elements not required by the program.

ARE exams are difficult, but if you study hard and prepare well, combined with your experience, IDP training, and/or college education, you should be able to pass all divisions and eventually be able to call yourself an architect.

14. English system (English or inch-pound units) vs. metric system (SI units)

This book is based on the English system or English units; the equivalent value in metric system or SI units follows in parentheses. All SI dimensions are in millimeters unless noted otherwise. The English or inch-pound units are based on the module used in the U.S. The SI units noted are simple conversions from the English units for information only and are not necessarily according to a metric module.

15. Codes and standards used in this book

We use the following codes and standards:
American Institute of Architects, Contract Documents, Washington, DC.
Canadian Construction Documents Committee, CCDC Standard Documents, latest edition, Ottawa.

16. Where can I find study materials on architectural history?

Every ARE exam may have a few questions related to architectural history. The following are some helpful links to FREE study materials on the topic:

http://www.essentialhumanities.net/arch.php
http://issuu.com/motimar/docs/history_synopsis?viewMode=magazine
http://www.ironwarrior.org/ARE/Materials_Methods/m_m_notes_2.pdf

Chapter Two

Site Planning & Design (SPD) Division

A. General Information

1. Exam Content

An architect should have the skills and knowledge of site planning and design, including economic, social and environmental issues, and practice and project management.

The exam content for the SPD division of the ARE includes:

1) **Principles (16% to 24%)**
   Evaluate and review the impact of design theory, historic precedent, human behavior in the selection of systems, materials, and methods on site design and construction:
   - **Architectural History and Theory:** Examine the site to determine the impact of the regional or local historic precedents, context, archeological and cultural value, forms, order and imagery, preservation of site features, and other historical issues.
   - **Implications of Design Decisions:** Assess the impact of site design decisions on schedules, costs, site utilization, basic engineering principles, expansion, or any other important site issue.
   - **Site Planning:** Select and evaluate a potential site; Assess and review the impact of zoning ordinances, building codes, and covenants, as well as physical, natural, and other constraints.
   - **Site Design and Design Principles:** Apply basic engineering principles and typical site design practices to site design. Apply the concepts of modeling and spatial visualization, form, shape, scale, line, pattern, symmetry, texture, color, contrast, mass, and acoustics to achieve programmatic site design goals.
   - **Adaptive Reuse of Buildings and/or Materials:** Evaluate the impact of adaptive reuse building and/or materials on site design.

2) **Environmental Issues (24% to 32%)**
   Review and assess environmental and site conditions, evaluate and apply materials, systems, and construction methods, integrate sustainable principles, and evaluate design impact on human behavior.
   - **Interpreting Existing Environmental and Site Conditions and Data:** Evaluate issues such as adjacencies, nearby amenities, view-sheds, topography, solar orientation, precipitation, prevailing winds, waterways, flooding, wildlife, vegetation, traffic, transportation, noise, infrastructure, and access. Consider the impact of environmental, climatological, and cultural constraints. Investigate environmental and site conditions.

- **Design Impact on Human Behavior:** Evaluate the impact of accessibility, orientation, exposure, lighting, color theory, acoustics, universal design, security, community cohesion, and neighborhood identity on human behavior.
- **Hazardous Materials and Conditions:** Review potential effect of hazardous materials conditions discovered during construction documents and construction phases and coordinate agency approvals, remediation/mitigation, and phasing plans.
- **Sustainable Design:** Utilize sustainable principles affecting energy consumption and the use of materials in site planning and design. Manage sustainable resources such as indigenous materials, low impact materials, and renewable and recyclable resources to minimize material consumption and waste.
- **New Material Technologies and Alternative Energy Systems:** Identify the impact of sustainable and environmental theories and principles on the design of building systems in site planning and design. Evaluate renewable and alternative energy sources and systems.

3) **Codes & Regulations (14% to 22%)**
Integrate zoning, specialty codes, building codes, and other regulatory requirements in site design and construction.
   - **Government and Regulatory Requirements and Permit Processes:** Code analysis & compliance, permitting processes
   - **Specialty Codes and Regulations including Accessibility Laws, Codes and Guidelines**: Fair Housing Act, ADAAG, seismic codes, life safety, and historic preservation requirements for the site planning portions of the construction documents, etc.

4) **Materials & Technology (22% to 30%)**
Review the influence design decisions have on the selection of materials, methods, and systems in site design and construction.
   - **Construction Details and Constructability:** Context, appropriate materials, longevity, and durability
   - **Construction Materials:** Selection of the appropriate site material
   - **Fixtures, Furniture, Equipment, and Finishes:** Selection of product/materials
   - **Product Selection and Availability:** Selection of products/materials, considering transportation, site location, and availability of site materials
   - **Thermal and Moisture Protection:** Applying principles of site drainage, moisture migration, water infiltration, and thermal expansion
   - **Natural and Artificial Lighting:** Site layout, building orientation, shape, daylight, solar control, and energy consumption on lighting requirements.
   - **Implications of Design Decisions:** Site layout, cost, engineering, schedule, life-cycle analysis, adaptability, materials selection, details, constructability, project schedule, cost, and regulatory agencies.

5) **Project & Practice Management (4% to 12%)**
Perform construction administration; assess construction sequencing, scheduling, cost, and risk management.

- **Construction Sequencing:** Phasing plans
- **Cost Estimating, Value Engineering, and Life cycle Costing:** Cost estimates, construction documents and construction administration.
- **Project Schedule Management:** Manage the scheduling of professional services
- **Risk Management:** Professional and general liabilities and risk management

The graphic vignettes include:

- **Site Grading Vignette:** You need to adjust a site's topographical characteristics per the regulatory requirements and the program.
- **Site Design Vignette:** You need to utilize general site planning principles to design a site, including parking, vehicular and pedestrian circulation, building placement, responding to environmental, programmatic, functional, and setback requirements.

2. Official exam guide and practice program for the SPD division

You need to read the official exam guide for the SPD division at least twice. Make sure you install the SPD division practice program on your computer and become very familiar with it. The real exam is VERY similar to the practice program.

You can download the official exam guide and practice program for the SPD division at the following link:
http://www.ncarb.org/en/ARE/Preparing-for-the-ARE.aspx

Note:
We suggest you study the vignettes in the official NCARB Study Guide first, and then study the official NCARB Multiple Choice (MC) sample questions, and then other study materials, and then come back to NCARB vignettes the NCARB (MC) questions again several days before the real ARE exam.

B. Important Documents and Publications for the SPD Division of the ARE Exam

SPD is one of the ARE divisions which is very hard to prepare for by simply reading a finite number of books within a short amount of time. This is because many of the questions in this division are based on work experience, and may also include questions from other ARE divisions.

We shall help you alleviate this problem by bringing your attention to some of the most common issues in architectural practice that relate to the SPD division.

The SPD division may include questions from a broad scope, but the questions **have to** be issues an **average** architect would encounter during normal architectural practice. NCARB may include some obscure issues, but not too many. Otherwise, the ARE tests would NOT be **legally defensible**.

Based on our research, the most important documents/publications for the SPD division of the ARE exam are:

1. ***The Architect's Handbook of Professional Practice* (AHPP)**
   Demkin, Joseph A., AIA, Executive Editor. *The Architect's Handbook of Professional Practice* (AHPP). The American Institute of Architects &Wiley, latest edition.

   This comprehensive book covers all aspects of architectural practice, and includes two CDs containing sample AIA contract documents. You may have a few real ARE SPD division questions based on this publication. Therefore, look through this book a few times and know some of the basic architectural practice elements. Read the related portions carefully, such as project delivery, project & practice management, and building codes and regulations.

2. ***Architectural Graphic Standards* (AGS)**
   Ramsey, Charles George, and John Ray Hoke Jr. *Architectural Graphic Standards.*
   The American Institute of Architects &Wiley, latest edition.

   There may be a few questions asking you to identify some of the basic graphic symbols. This is a good book to skim through.

3. **Access Board, *ADAAG Manual: A Guide to the Americans with Disabilities Accessibility Guidelines***. East Providence, RI: BNI Building News. ADA Standards for Accessible Design are available at:

   http://www.ada.gov/

   AND
   http://www.access-board.gov/guidelines-and-standards/buildings-and-sites/about-the-ada-standards/ada-standards

4. **The following documents from the EPA:**
   Wetland Overview:
   http://water.epa.gov/type/wetlands/upload/2005_01_12_wetlands_overview.pdf

   Developing your Stormwater Pollution Prevention Plan: A guide for Construction Sites:
   http://www.epa.gov/npdes/pubs/sw_swppp_guide.pdf

   You may have about 20 questions on environmental issues. These documents will help you. You should **read** them **at least twice** to become familiar with them.

5. ***LEED Green Associate Exam Guide***: *A Must-Have for the LEED Green Associate Exam: Comprehensive Study Materials, Sample Questions, Mock Exam, Green Building LEED Certification, and Sustainability*. ArchiteG, Inc., latest edition.

This book is a good introduction to green buildings, environmental issues, and the LEED building rating systems. You just need to look through it, and focus on the generic information on green buildings and sustainability since you are not taking a LEED exam.

6. **American Institute of Architects (AIA) Documents**
You may have several real ARE SPD division questions based on AIA documents. The questions will deal with the frontend stuff of a project, i.e., issues related to site planning & design, project management & practice like scope of work and contract, and NOT the backend stuff, i.e., constructions administration issues like job site observation, submittals, and shop drawings.

Reading the summary of the AIA Documents is NOT adequate preparation. You need to read the complete text. Fortunately, you do NOT have to read all the available AIA documents.

Three possible study solutions are as follows:

a. Buy ONLY the AIA documents you need from your local AIA office. The AIA documents listed below are important, especially those in **bold** font, read them at least three times. You may have **some** real ARE SPD division questions based on the following AIA documents listed in **bold** font:

- **A101–2007, Standard Form of Agreement Between Owner and Contractor where the basis of payment is a Stipulated Sum (CCDC Document 2)**
- **A201–2007, General Conditions of the Contract for Construction**
- A503–2007, Guide for Supplementary Conditions
- A701–1997, Instructions to Bidders (CCDC Document 23)
- **B101–2007 (Former B141–1997), Standard Form of Agreement Between Owner and Architect (RAIC Document 6)**

  You can find FREE sample forms and commentaries for AIA documents A201 & B101 at the following link:
  http://www.aia.org/contractdocs/aiab081438

- **C401–2007 (Former C141–1997), Standard Form of Agreement Between Architect and Consultant**

  There are some major changes between C401-2007 and the older version C141–1997.

  However, a FREE version of sample C141–1997 is available at:
  https://app.ncarb.org/are/StudyAids/_C141.pdf

- G701–2001, Change Order
- G702–1992, Application and Certificate for Payment
- G704–2000, Certificate of Substantial Completion

AIA updates their documents roughly every 10 years. Although, please note that AIA does NOT update all available documents at the same time. For example, A701–1997, Instructions to Bidders (CCDC Document 23) is still the most current form.

See the AIA documents price list at the following link for the latest edition of AIA documents:
http://www.aia.org/aiaucmp/groups/aia/documents/pdf/aias076346.pdf

b. Buy AHPP, which has a CD including the sample AIA documents. AHPP itself is also a very important publication for the SPD division.

c. After you register with NCARB for the ARE exams and log in to their site, you have FREE access to the AIA documents online.

7. ***Fundamentals of Building Construction, Materials, and Methods.*** Latest Edition by Edward Allen. John Wiley & Sons, latest edition

   You should focus on the information on soils and foundations.

8. **Historic Preservation documents**
   Based on examinee feedback, there are a few questions regarding historic preservation in the SPD division exam.

   You should **read** the following documents **at least twice** to become familiar with them. Pay special attention to information in italics and the shaded areas of the PDF file:
   - The two FREE PDF files are *The Secretary of the Interior's Standards for the Treatment of Historic Properties with Guidelines for Preserving, Rehabilitating Restoring & Reconstructing Historic Buildings* and *The Secretary of the Interior's Standards for Rehabilitation & Illustrated Guidelines for Rehabilitating Historic Buildings- Standards* available at:
     http://www.ironwarrior.org/ARE/Historic_Preservation/

     AND
     http://www.nps.gov/hps/tps/tax/rhb/

   You should **look through** the following document and become familiar with it:
   - ***National Historic Preservation Act (NHPA)*** at the following link:
     http://www.gsa.gov/portal/content/104441

9. **Construction Specifications Institute (CSI) MasterFormat & *Building Construction***
   Become familiar with the new 6-digit CSI Construction Specifications Institute (CSI) MasterFormat as there may be a few questions based on this publication. Make sure you know which items/materials belong to which CSI MasterFormat specification section, and memorize the major section names and related numbers. For example, Division 9 is Finishes, and Division 5 is Metal, etc. Another one of my books, *Building Construction*, has detailed discussions on CSI MasterFormat specification sections.

Mnemonics for the 2004 CSI MasterFormat

The following is a good mnemonic, which relates to the 2004 CSI MasterFormat division names. Bold font signals the gaps in the numbering sequence.

This tool can save you lots of time: if you can remember the four sentences below, you can easily memorize the order of the 2004 CSI MasterFormat divisions. The number sequencing is a bit more difficult, but can be mastered if you remember the five bold words and numbers that are not sequential. Memorizing this material will not only help you in several divisions of the ARE, but also in real architectural practice

Mnemonics (pay attention to the underlined letters):
Good students can memorize material when teachers order.
F students earn F's simply 'cause **forgetting** principles have **an** effect. (21 and 25)
C students **end** everyday understanding things without memorizing. (31)
Please make professional pollution prevention inventions **every day**. (40 and 48)

1-Good.................................. General Requirements
2-Students.............................. (Site) now Existing Conditions
3-Can.....................................Concrete
4-Memorize............................Masonry
5-Material ....Metals
6-When..................................Woods and Plastics
7-Teachers.............................Thermal and Moisture
8-Order..................................Openings

9-F...Finishes
10-Students............................ Specialties
11-Earn.................................Equipment
12-F's....................................Furnishings
13-Simply..............................Special Construction
14-'Cause...............................Conveying
21-Forgetting .... Fire
22-Principles.......................... Plumbing
23-Have................................ HVAC
25-An.................................... Automation
26-Effect............................... Electric

27-C...................................... Communication
28-Students............................ Safety & Security
31-End.................................. Earthwork
32-Everyday...........................Exterior
33-Understanding ....Utilities
34-Things..............................Transportation
35-Without Memorizing........ Waterways and Marine

40-Please................................Process Integration
41-Make................................ Material Processing and Handling Equipment
42-Professional...................... Process Heating, Cooling, and Drying Equipment
43-Pollution.......................... Process Gas and Liquid Handling, Purification and Storage Equipment
44-Prevention........................ Pollution Control Equipment
45-Inventions........................ Industry-Specific Manufacturing Equipment
48-Everyday.........................Electrical Power Generation

Note:
There are 49 CSI divisions. The "missing" divisions are those "reserved for future expansion" by CSI. They are intentionally omitted from the list.

C. Overall Strategies and Tips for NCARB Graphic Vignettes

1. Overall strategies

To most candidates, the Multiple Choice (MC) portion of an ARE division is harder than the graphic vignette portion. Some of the MC questions are based on experience and you do NOT have a set of fixed study materials for them. You WILL make some mistakes on the MC questions no matter how hard you study.

On the other hand, the graphic vignettes are relatively easier, and there are good ways to prepare for them. You should really take the time to study and practice the NCARB graphic software well, and try to **nail all the graphic vignettes** perfectly. This way, you will have a better chance to pass even if you answer some MC questions incorrectly.

Tips: *Most people do poorly on the MC portion of the SPD division, especially those who do NOT have a lot of working experience, but curiously not too many people fail because of the MC portion. Most people fail the ENTIRE SPD section because they have made* ***one*** *fatal mistake on the graphic vignette section. So, practice the NCARB SPD practice program graphic software and make sure you absolutely NAIL the vignette section. This is a key for you to pass.*

The official NCARB SPD exam guide gives a passing and failing solution the sample vignette, but it does NOT show you the step-by-step details.

We are going to fill in the blanks here and offer you step-by-step instructions, command by command.

You really need to spend time practicing to become VERY familiar with NCARB's graphic software. This is because all ARE graphic vignettes are timed, and you do NOT have the luxury to think about how to use the software during the exam. Otherwise, you may NOT be able to finish your solution on time.

The following solution is based on the official NCARB SPD practice program for the **ARE 4.0**. Future versions of ARE may have some minor changes, but the principles and fundamental elements should be the same. The official NCARB SPD practice program has not changed much since its introduction and the earlier versions are VERY similar to, if not exactly the same as, the current ARE 4.0. The actual graphic vignette of the SPD division should be VERY, VERY similar to the practice one on NCARB's website.

2. **Tips**

1) You need to install the NCARB SPD practice program, and become familiar with it. I am NOT going to repeat the vignette description and requirements here since they are already written in the NCARB practice program.

   See the following link for a FREE download of the NCARB practice program: http://www.ncarb.org/ARE/Preparing-for-the-ARE.aspx

2) Review the general test directions, vignette directions, program, and tips carefully.
3) Press the space bar to go to the work screen.
4) Read the program and codes in the NCARB Exam Guide several times the week before your exam. Become VERY familiar with this material, and you will be able to read the problem requirements MUCH faster during the real exam because you can immediately identify which criteria are different from the practice exam.

3. **A Step-by-step solution to the official NCARB practice program graphic vignette: Site Grading**

1) **Overall concept:**

- The Locomotive Display needs to be level: This means you need to have a contour line of the same elevation wrap around the Display.
- Regrade the site so that water will flow around and away from the Locomotive Display: This means you need to create 2 swales and wrap them around the Display.
- The slope of the regraded portions of the site shall be at least 2% and no more than 20%: Since the vertical elevation difference between the adjacent contour lines is 1', this means the horizontal distance between the adjacent contour lines has to be between 5' and 50'.

  Note:
  Slope = (difference of adjacent contour elevations/ horizontal distance between the adjacent contour lines)

  Horizontal distance between the adjacent contour lines = (difference of adjacent contour elevations/slope)

  1'/20% = 5'
  1'/2% = 50'

2) Use **Draw>Locomotive Display** to draw a Locomotive Display within the building limit line (Figure 2.1). Leave about the same space between the Display and the building limit line, and between the Display and the existing rocks.

3) Use **Sketch>Circle** to pre-draw a dozen 5'-diameter (OR 2'-6"-radius) circles for checking the maximum slope later. Once you draw one 5'-diameter circle, and then each time you click, the software will automatically draw another 5'-diameter circle (Figure 2.2).

4) Use **Sketch** > **Line** to draw the centerlines of the two swales. The two swales should start on the high side of the Locomotive Display, and wrap around the sides of the Display to divert the water (Figure 2.3).

5) Click on **Move, Adjust**, and the contour lines will become highlighted (Figure 2.4).

6) Click on the contour lines (NOT necessarily right on the small squares) to adjust them to form the two swales (Figure 2.5).

   Note:
   - *Since the existing 106' contour line is the lowest existing contour line that hits the Display, we select it as the contour line that wraps around the Locomotive Display.*
   - *Start by adjusting the 106' contour line that wraps around the Locomotive Display, and then adjust other contour lines accordingly.*
   - *The direction of the water flow is opposite to the "arrow" formed by a swale's contour lines.*
   - *Use the pre-drawn 5'-diameter circles to check and make sure the horizontal distance between the adjacent contour lines is larger than 5' (less than 20% slope). Adjust the contour lines if necessary (Figure 2.6).*
   - *Use sketch line and id tool to check and make sure the horizontal distance between the adjacent contour lines is less than 50' (larger than 2% slope). Adjust the contour lines if necessary (Figure 2.7).*
   - *The slope between the Locomotive Display and the contour line that wraps around the Locomotive Display is not subject to the maximum slope requirement, so place the contour line as close to the Locomotive Display as possible. Adjust the contour lines if necessary (Figure 2.8).*
   - *Make sure the contours around the smoke stack, the rock and the trees are NOT disturbed.*

7) Use Zoom to zoom out, and click on the **Set Elevation** button on the left-hand side menu. A dialog box appears. Use the **up and down arrow** to set the elevation to 106'-6". This is 6" higher than the 106' contour line that wrap around the Locomotive Display (Figure 2.9).

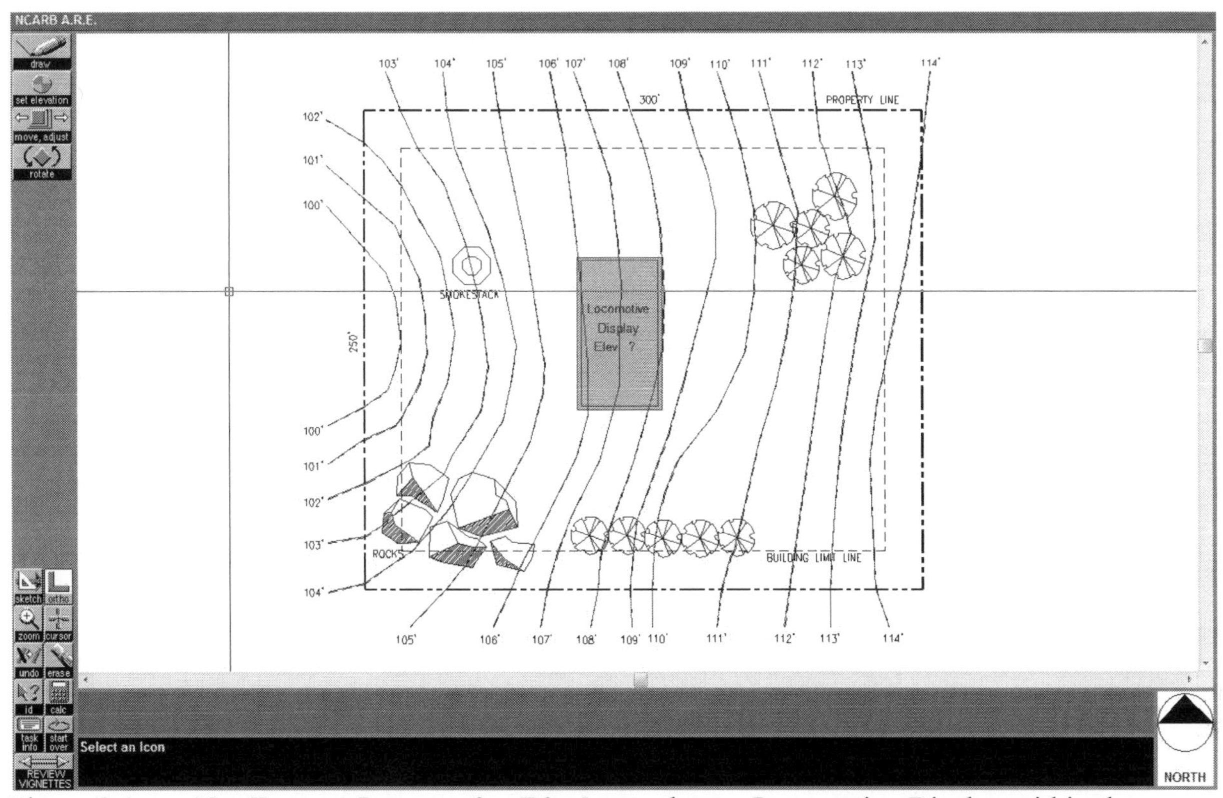

Figure 2.1 Use **Draw > Locomotive Display** to draw a Locomotive Display within the building limit line.

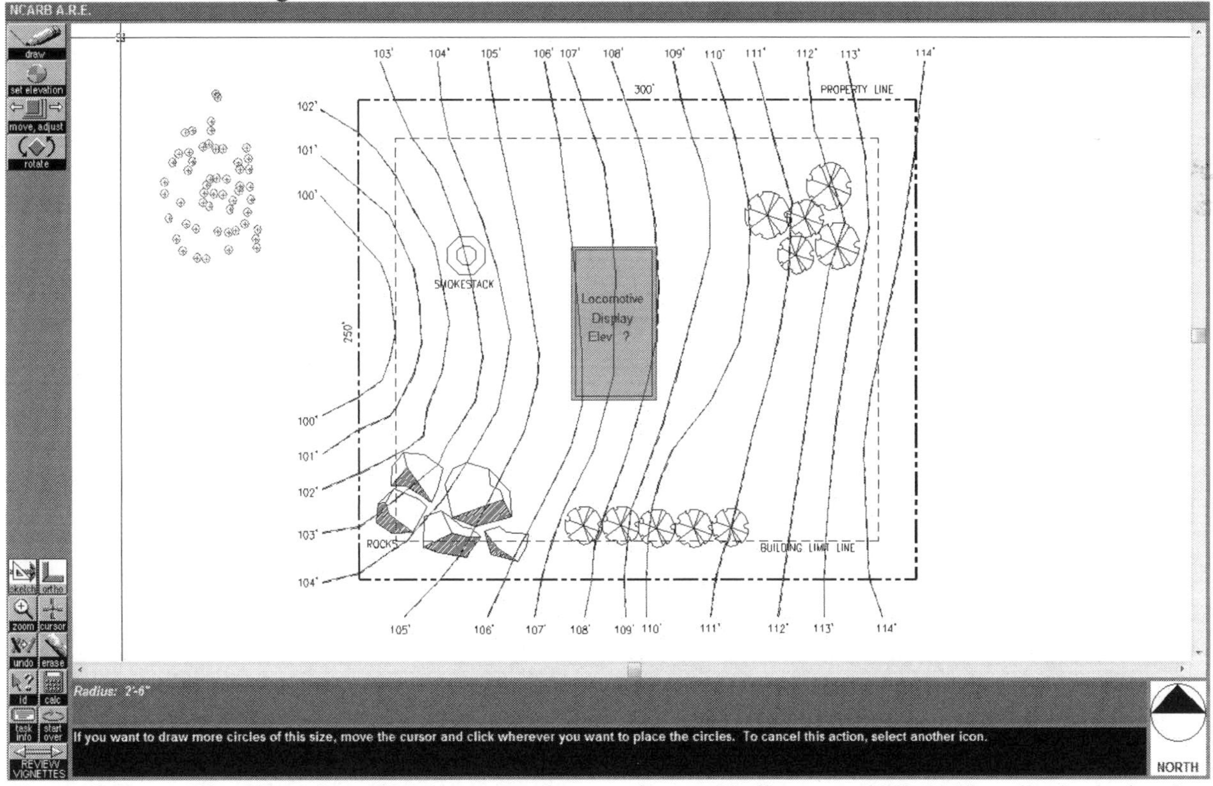

Figure 2.2 Use **Sketch > Circle** to pre-draw a dozen 5'-diameter (OR 2'6"-radius) circles for checking the maximum slope later.

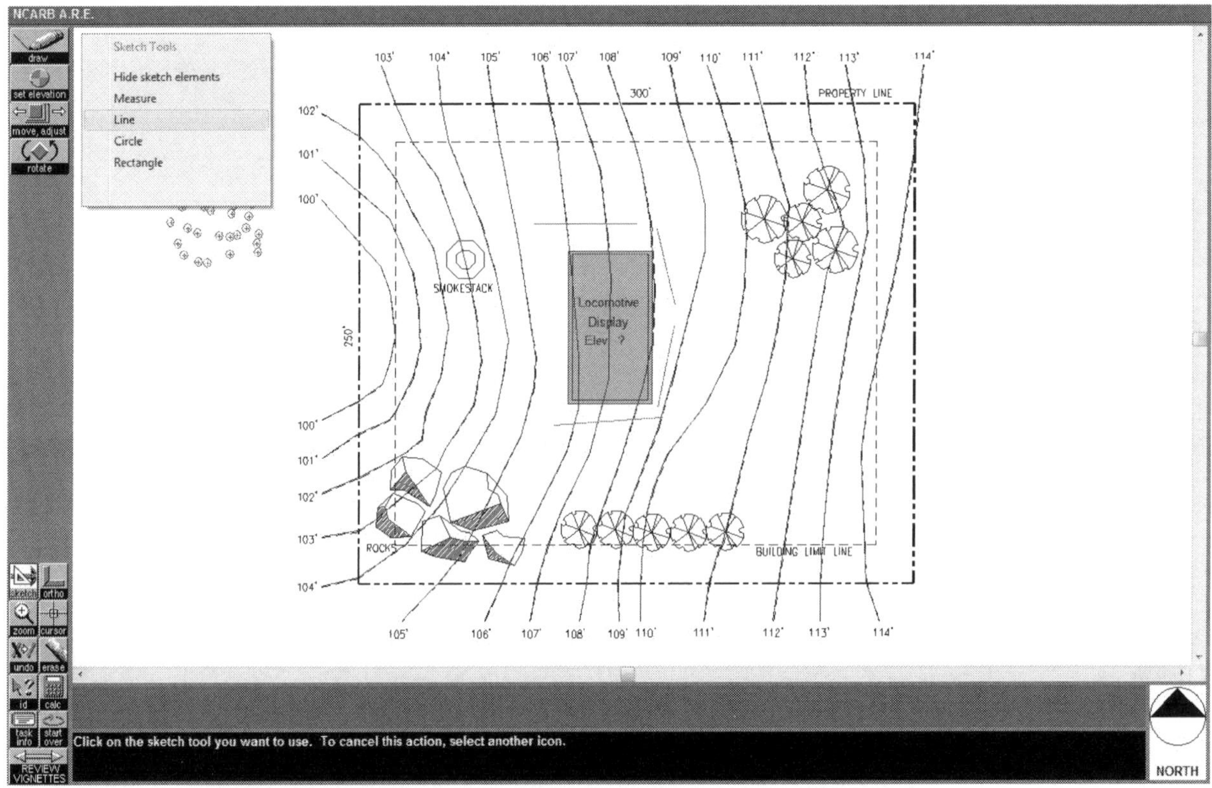

Figure 2.3 Use **Sketch > Line** to draw the centerlines of the two swales.

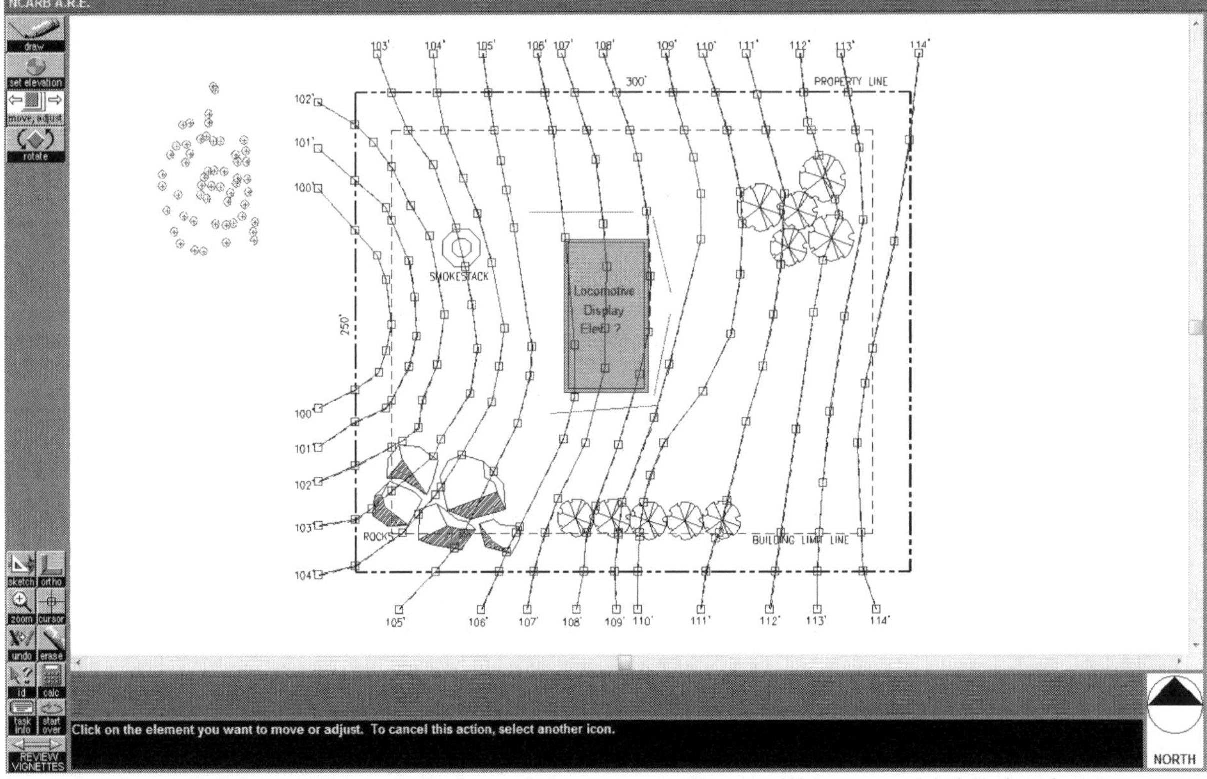

Figure 2.4 Click on **Move, Adjust**, and the contour lines will become highlighted.

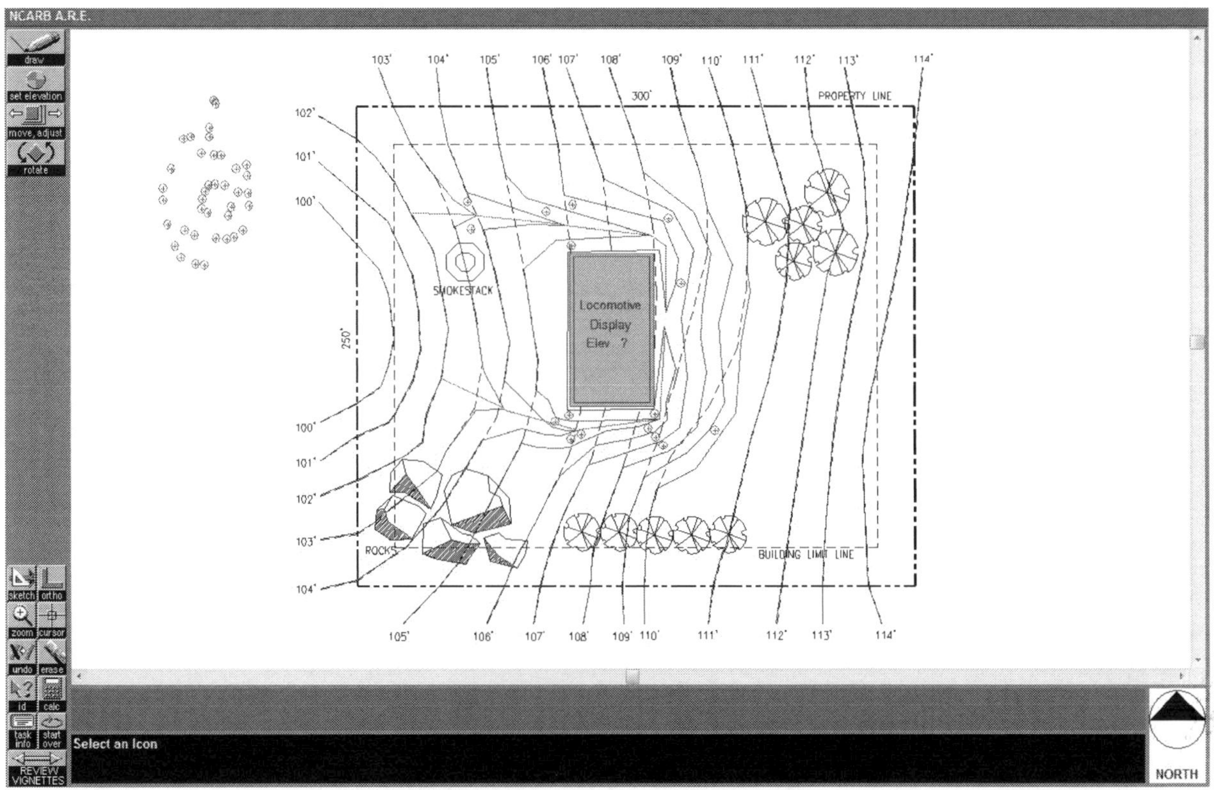

Figure 2.5 Click on the contour lines (NOT necessarily right on the small squares) to adjust them to form the two swales.

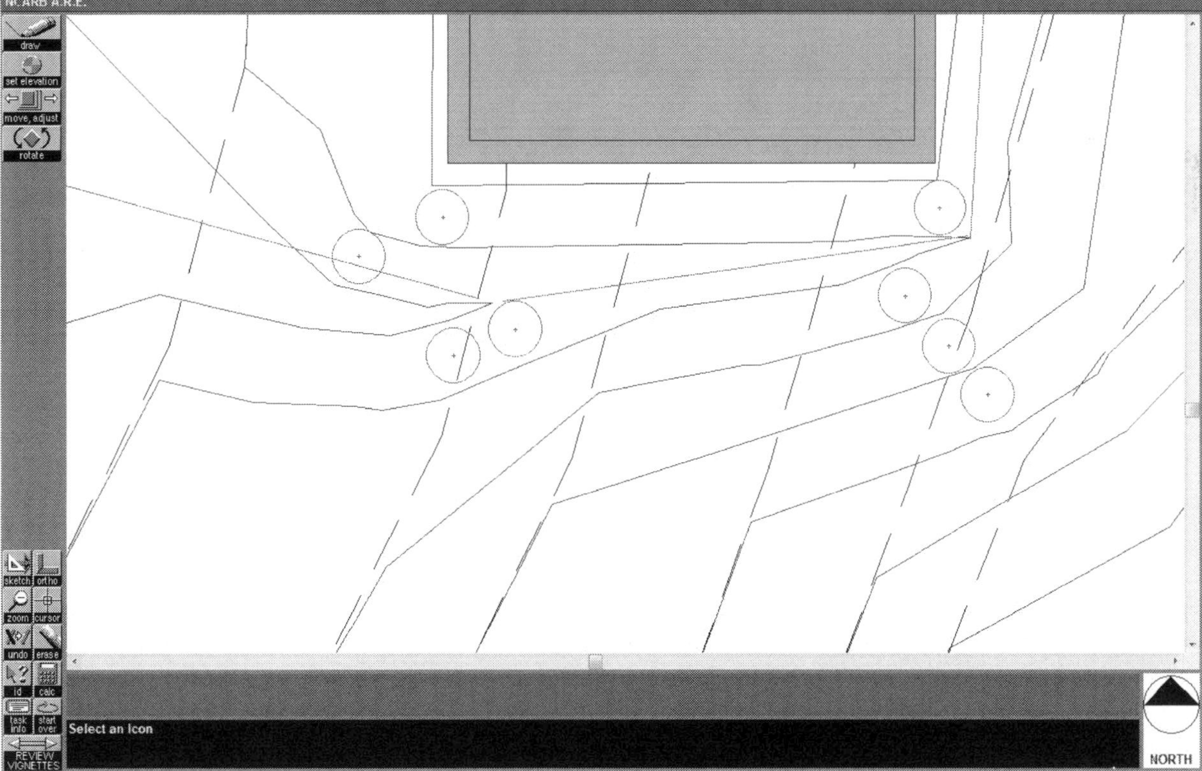

Figure 2.6 Use the pre-drawn 5'-diameter circles to check and make sure the horizontal distance between the adjacent contour lines is larger than 5'.

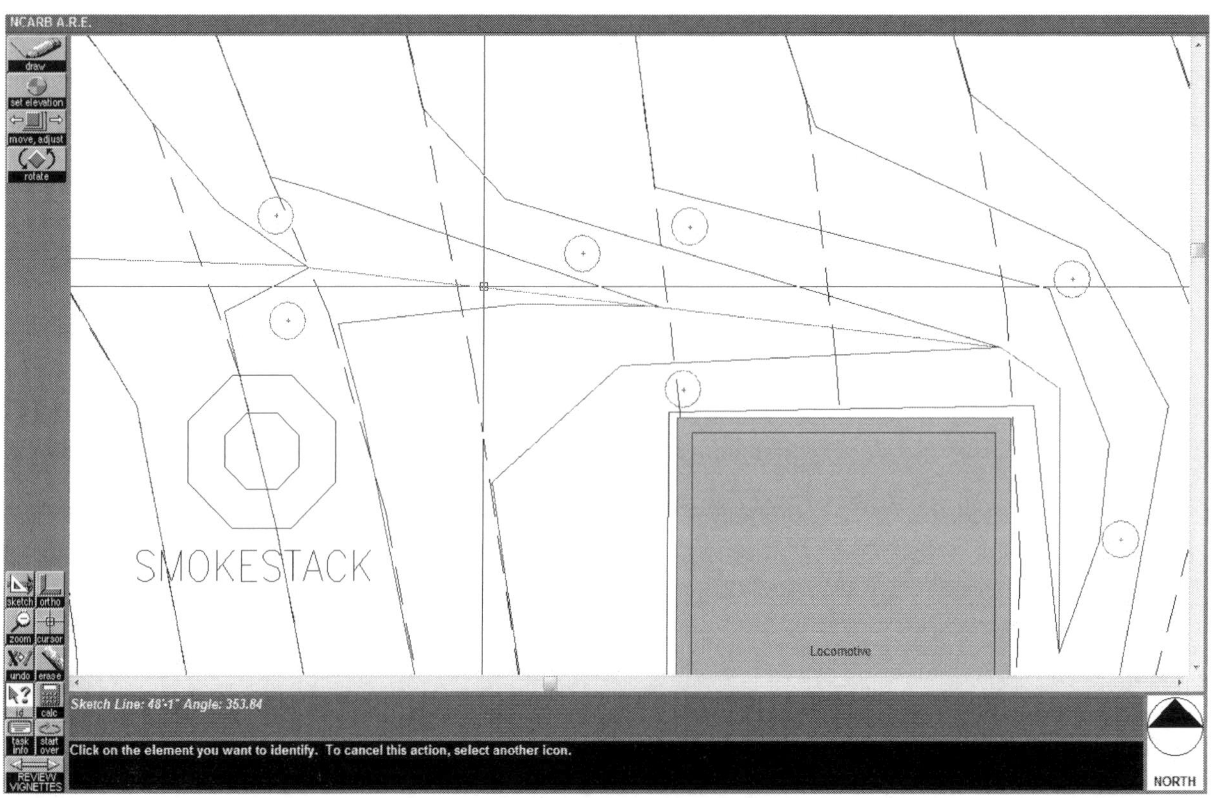

Figure 2.7 Use sketch line and id tool to check and make sure the horizontal distance between the adjacent contour lines is less than 50'.

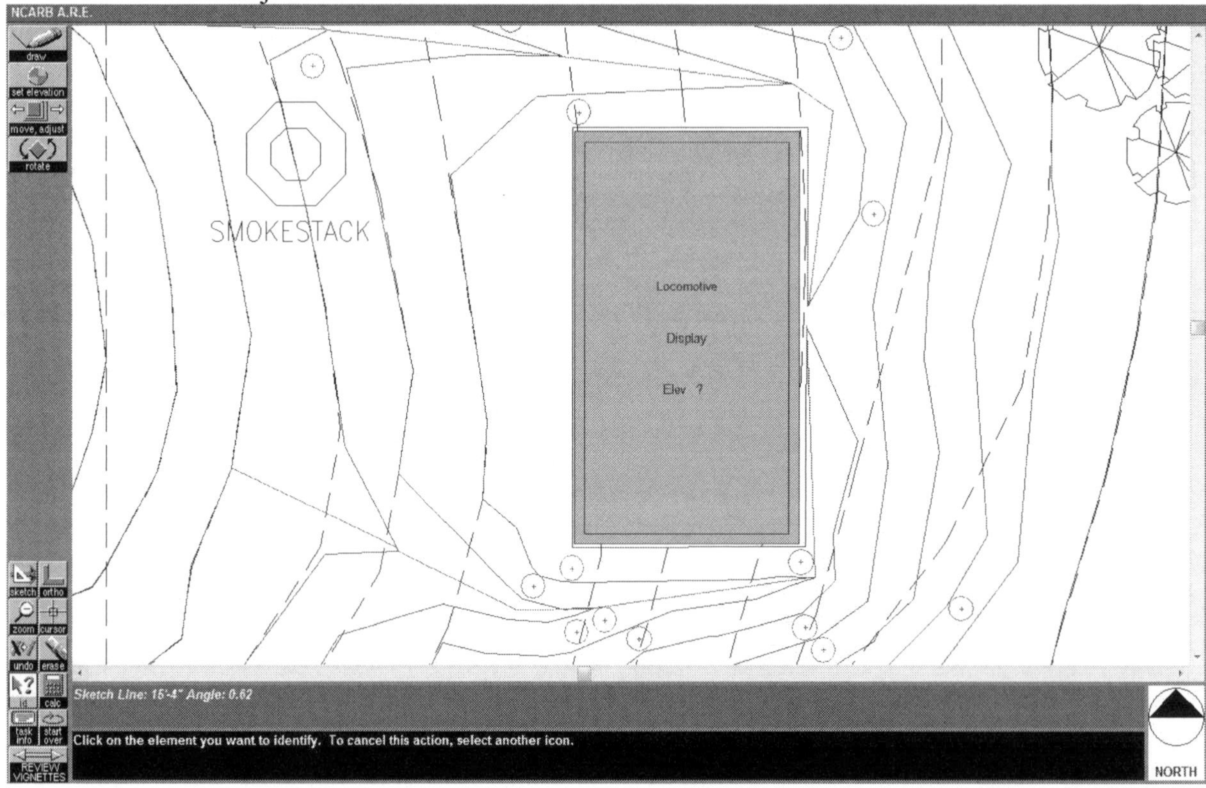

Figure 2.8 The slope between the Locomotive Display and the contour line that wraps around the Locomotive Display is not subject to the maximum slope requirement.

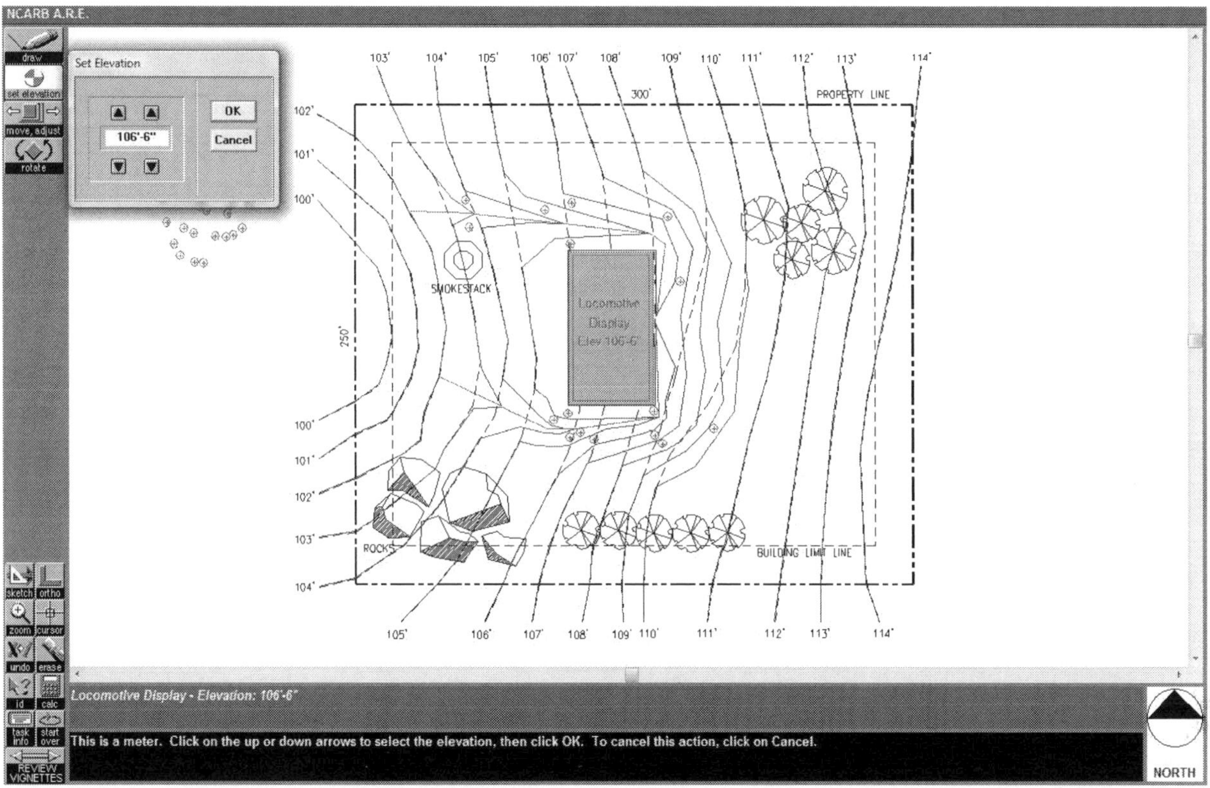

Figure 2.9 Use the **up and down arrow** to set the elevation to 106'-6". This is 6" higher than the 106' contour line that wraps around the Locomotive Display.

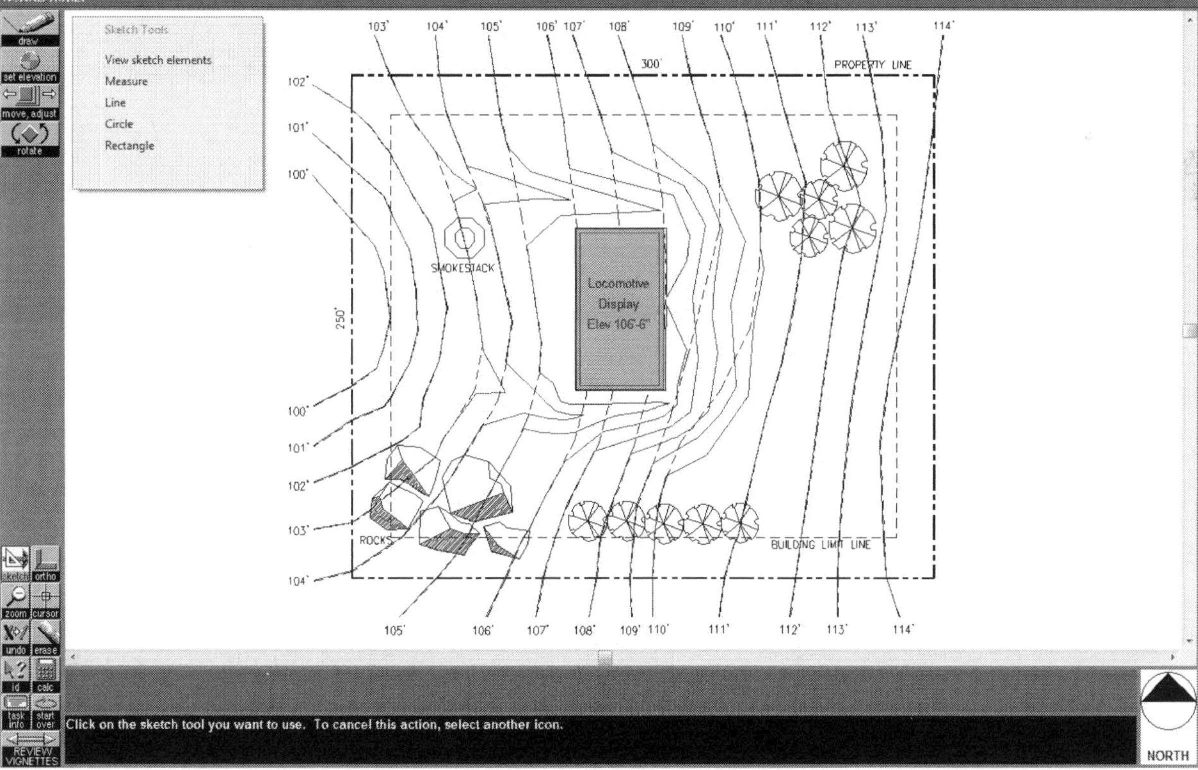

Figure 2.10 Click on the **Sketch > Hide sketch elements**, and the sketch lines and circles will disappear. This is your final solution.

8) Click on the **Sketch** > **Hide sketch elements**, and the sketch lines and circles will disappear. This is your final solution (Figure 2.10).

Note:
Another alternate solution is to select the existing 107' contour line that hits the middle of the Display as the contour line that wraps around the Locomotive Display. This way, the elevation of the Display will be 107'-6", the ***cut and fill*** *amount of the soils will be roughly equal, and you can reduce labor of earthwork and related construction cost.*

4. Items that you need to pay special attention to

Several items that you need to pay special attention to:

1) This vignette seems to be very simple, but you still need to check all the program requirements and make sure you comply with all of them.
2) Select a proper contour line as the contour line that wraps around the display area to create a level area.
3) The direction of the water flow is opposite to the "arrow" formed by a swale's contour lines.
4) Use the pre-drawn 5'-diameter circles to check and make sure the horizontal distance between the adjacent contour lines is larger than 5' (less than 20% slope). Adjust the contour lines if necessary.
5) Use sketch line and id tool to check and make sure the horizontal distance between the adjacent contour lines is less than 50' (larger than 2% slope). Adjust the contour lines if necessary.
6) The slope between the Locomotive Display and the contour line that wraps around the Locomotive Display is not subject to the maximum slope requirement, so place the contour line as close to the Locomotive Display as possible. Adjust the contour lines if necessary.
7) Make sure the contours around the smoke stack, the rock and the trees are NOT disturbed.

5. A Step-by-step solution to the official NCARB practice program graphic vignette: Site Design

1) **Overall concept:**

- The Site Design vignette has a very complicated program. If you simply read the program several times, you may NOT cover all the program requirements when you layout the site design. It is also VERY hard to understand and summarize the program requirements by simply reading it multiple times.
- One effective way of digesting the program requirements is to do a simple **Bubble Diagram**: use the paper and pencil provided by the test center to draw a simple **bubble diagram** to show the relationships between the site elements (Figure 2.11). This is the HARDEST part of the vignette.
  *Note: A **bubble diagram** is the MOST important tool for passing Site Design vignette.*

  - The bubble diagram does NOT have to be pretty and does NOT have to be to scale. The key is to do the bubble diagram quickly and turn the program requirements into a graphic in the shortest time. The bubble diagram is for YOU. As long as YOU can read it, it is good enough. Architects love graphic thinking. A bubble diagram can be a great tool.
  - A **bubble diagram** can be a few simple hand-drawn circles: A circle with a label such as OT represents the Office Tower; R represents the Restaurant; PP represents the Pedestrian Plaza; WW represents a walkway; an arrow between two circle represents direct access required between the two elements; an arrow with a number above it represents a clearance distance between the two elements; a line connecting two elements indicates that they need to be placed close to each other; an eye symbol represents an element should be visible from a certain location (Figure 2.11), etc.

2) The program requires us to provide only one 24 ft wide curb cut located no closer than 120 ft from the intersection of the centerlines of the two existing public streets. Use **Sketch** > **Circle** to draw a circle with a 120-ft radius and with its center at the intersection of the centerlines of the two existing public streets, the 24 ft wide curb cut must be outside of this sketch circle (Figure 2.12).

3) Provide a 30 ft setback from the Pond for all construction or built improvements. Use **Zoom** to zoom into the pond area. Use **Sketch** > **Circle** to draw a number of circles with 15ft radii defining the 30 ft setback from the pond (Figure 2.13).

   Note:
   - *The radius of the circle displays on the lower-left-hand corner of the screen when drawn. After you draw the first circle, the radius for rest of the circles will stay at 15ft, and you just need to click on the screen to place them.*
   - *A circle with a 15-ft radius has a 30 ft diameter, which equals the required 30 ft setback from the pond. If you use circles with 30 ft radii, your set back will be 60*

ft, and you will lose a lot of time and may NOT be able to come up with a passing solution.

4) Locate the 5-story, 60 ft high Office Tower close to the Pond, and the main entrance shall be visible from Bentley Avenue. Use **Zoom** to zoom out. Use **Draw** > **Office Tower** to draw the office tower near the pond with its main entrance facing Bentley Avenue (Figure 2.14).

5) Use **Draw** > **Restaurant** to draw the restaurant (Figure 2.15).

6) The main entrance of the Restaurant shall receive the noonday summer sun: Assume a 45° solar altitude angle. This means the main entrance of the Restaurant shall face south and need to avoid the shadow of the 60 ft high Office tower and the existing 35ft high shopping center. Click on **Rotate**, click on the restaurant, and click on **Rotate** again to rotate the restaurant. Use **Move, Adjust** to move it up to make room for the Pedestrian Plaza (Figure 2.16).

   Note: *Pay attention to the legend on the lower-right hand corner of the screen. The main entrance is marked with a* ***solid-filled*** *triangle, and the service entrance is marked with a* ***regular*** *triangle.*

7) Buildings must be separated by a minimum of 20 ft. Use **Sketch** > **Line** to check and make sure the distance between the Office Tower and the Restaurant is larger than 20 feet. Check to make sure the distance between the Office Tower and the Existing Shopping Center is larger than 20 feet also.

8) Click on **Ortho** to turn on the Ortho mode. Use **Zoom** to zoom in. Use **Draw** > **Pedestrian Plaza** to draw the Pedestrian Plaza (Figure 2.17).

9) Click on **id?** then click on the Pedestrian Plaza. The square footage (9,327 sf) of the Pedestrian Plaza appears on the lower-left-hand corner of the screen. The actual square footage of the Pedestrian Plaza is too large. Use **Move, Adjust** to adjust the square footage to as close to 8,000 sf, the required square footage, as possible (Figure 2.18). We are able to adjust the square footage of the Pedestrian Plaza to 7,995 sf. This is very close to 8,000 sf.

   Note: *In any case, the Pedestrian Plaza should NOT be 10% larger or smaller than the required size.*

10) Locate the universally accessible parking spaces within 100 ft of the main entrance of the Office Tower. Use **Sketch** > **Circle** to draw a circle with a 100-ft radius and with its center at the main entrance of the Office Tower (Figure 2.19).

11) 3 universally accessible (12 ft x 18 ft) parking spaces are required. Click on **Draw** > **Handicap Spaces**, a dialogue box will pop up. Click on the **down arrow** to set the number of Handicap Spaces to 3 (Figure 2.20).

12) Click on **OK**, and then Click on **Ortho** to turn on the Ortho mode. The Handicap Spaces are drawn as 3 point rectangles. Set the over widths of the Handicap Spaces to 36' (12'x3=36'), and the depth of the Handicap Spaces to 18'. The dimensions of the Handicap Spaces appear on the lower-left-hand corner of the screen (Figure 2.21).

    Note: *You should ONLY use **Draw > Handicap Spaces** to draw the handicap spaces. Do NOT use **Draw > Standard Spaces** to draw the handicap spaces.*

13) Drives, traffic aisles, and parking spaces shall be no closer than 5 ft to a building. Use **Draw > Sketch Rectangle** to draw a 5' wide sketch rectangles to set the 5 ft clearance for the Office Tower and the Restaurant (Figure 2.22).

14) Use **Rotate** to rotate the 3 Handicap Spaces. Use **Move, Adjust** to move them closer to the Office Tower (Figure 2.23).

15) The intersection of the access drive with the street must be perpendicular to the street for at least the first 20 ft of the drive. Use **Draw > Driveway** to draw a drive from Bentley Avenue, to the left of the utility easement, and at least 5' from the Restaurant (Figure 2.24).

    Note: *Use **Draw > Sketch Rectangle** to draw a 24' wide sketch rectangles to assist you to locate the driveway if necessary.*

16) Use **Draw > Driveway** to draw a service drive to the service entrance of the Restaurant (Figure 2.25).

17) 30 standard (9 ft x 18 ft) parking spaces are required. Use **Draw > Standard Spaces** to draw 30 standard spaces. Pay attention to the following requirements (Figure 2.26):
    - Drive-through circulation is required.
    - Dead-end parking is prohibited.
    - Parking along the service drive is prohibited.

    Note: *Use Draw > **Sketch Rectangle** to draw a 24' wide sketch rectangles to assist you to locate the driveways and parking spaces if necessary.*

    If a space is too narrow for placing a driveway, you can use move group to move the other elements to make room for the driveway.

18) Use **Draw > Driveway** to draw the remaining driveways (Figure 2.27).

    Note:
    - *The key to draw, align or join different sections of the driveway is to align the centerlines of the driveways.*
    - *Use Move, Adjust to adjust the length of the driveways if necessary.*
    - *Sometimes it may be easier to erase a driveway and then redraw it instead of trying to adjust it.*

19) Use **Draw >Walkway** to draw a sidewalk from the Handicap Spaces to the Pedestrian Plaza, and a sidewalk from the Pedestrian Plaza to the Public Walk along Bentley Avenue (Figure 2.28).

20) The service entrance is NOT facing the Pedestrian Plaza. It is already blocked by the Restaurant itself from the Pedestrian Plaza (Figure 2.29).

    Note: *The view of the service entrance (NOT the entire service entrance and loading area) on the Restaurant shall be blocked from the Pedestrian Plaza by buildings and/or trees.*

21) The Pedestrian Plaza shall be blocked from the prevailing winter winds by buildings and/or trees. Pay attention to the *arrow marked with "winds"* on the upper-right hand corner of the screen. Use **Sketch > Line** to draw a few sketch lines parallel to the direction of wind to assist you to place coniferous trees to block the Pedestrian Plaza from the prevailing winter winds.

    Note: *The sketch lines parallel to the direction of winds are very valuable tool for you to layout the coniferous trees.*

22) Use **Draw > Coniferous Tree** to draw several coniferous trees to block the winds (Figure 2.30).

    Note:
    - *Do NOT overlap trees.*
    - *Make sure the main entrance of the Office Tower is visible from Bentley Avenue when adding trees.*
    - *Pay attention to the elevation of trees: Deciduous trees will allow views under the tree crown; coniferous trees have a triangle shape, and may still allow view or winds to pass through if you simply place them side by side.*
    - *Do NOT draw more trees than necessary because you may accidentally block some other elements or violate another program requirements.*

23) Use **Sketch > Hide Sketch Elements** to hide sketch elements. We notice the bank of standard parking spaces along the south side of the Pedestrian Plaza is at a wrong orientation (Figure 2.31).

24) Use **Rotate** to rotate the parking spaces 180 degrees, and use **Move Group** to move them to the right location. We notice the bank of standard parking spaces along the east side of the site is not completely connected to the driveway. Use **Move Group** to move them to the right location. Use **Sketch > Hide Sketch Elements** to hide sketch elements (Figure 2.32).

25) Click on **Check**, all the trees that overlap with sidewalk, parking space or other elements will be highlighted (Figure 2.33). We have 7 highlighted trees, but only 6 of them need to be cut as allowed by the program.

Note:

- *If a tree overlaps a parking space or sidewalk, it may not need to be cut. By reviewing the elevation of the trees provided by the NCARB program, you notice that a small portion of the crown of a deciduous tree can overlap a parking space or sidewalk. On the other hand, if the crown of a coniferous tree overlaps a parking space or sidewalk, the tree will be cut.*
- *If you are NOT comfortable with having 7 highlighted trees, you can adjust the walkway between the Pedestrian Plaza and the public walkway (Figure 2.34) or the parking spaces so that only 6 trees are highlighted. You can reduce the number of the parking spaces along the bottom bank of parking spaces, and add the same number of parking spaces along the east bank of parking spaces.*

26) After finishing your solution, re-read the program again carefully to double check and make sure you cover every requirement.

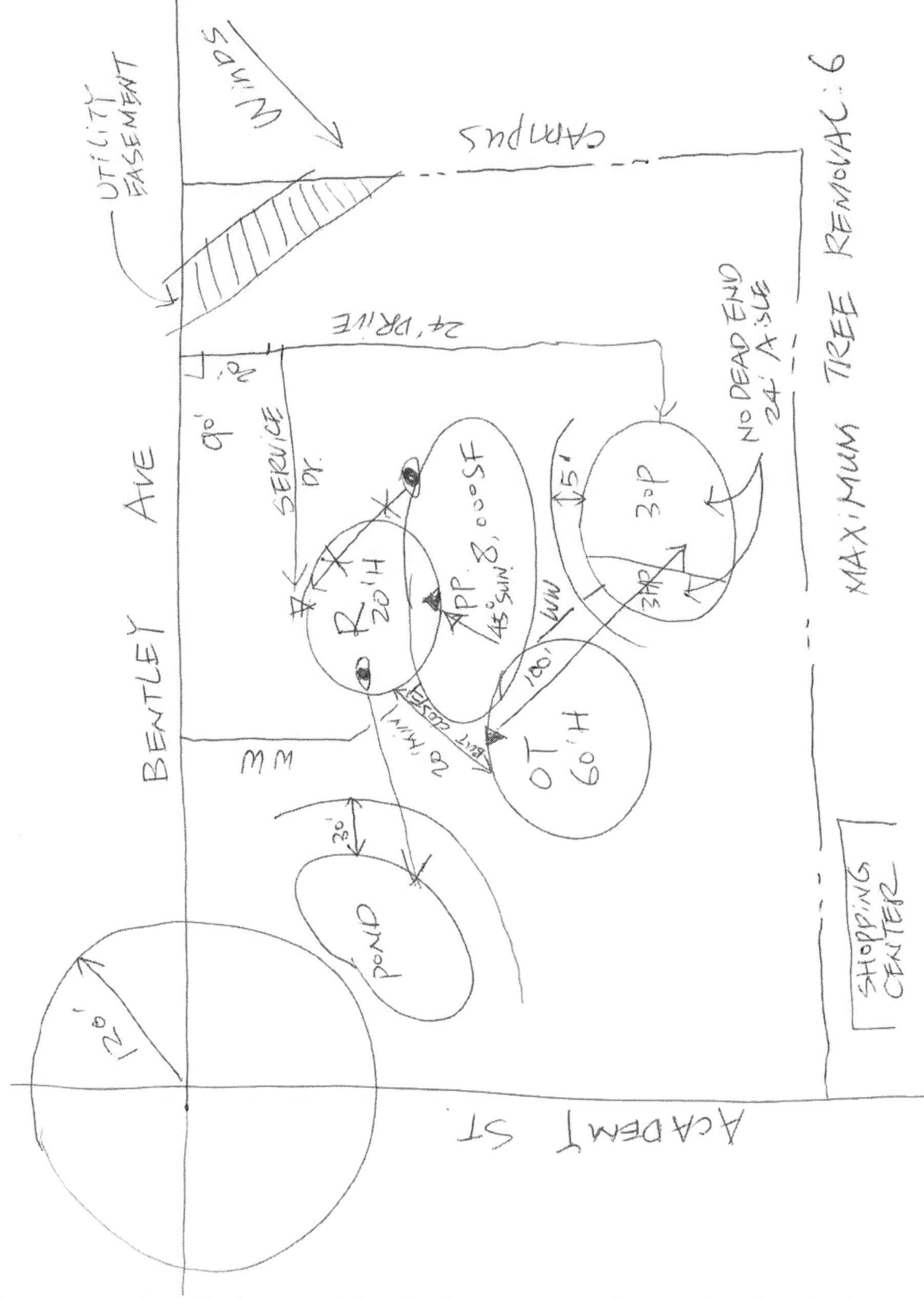

Figure 2.11 One effective way of digesting the program requirements is to do a simple **Bubble Diagram**.

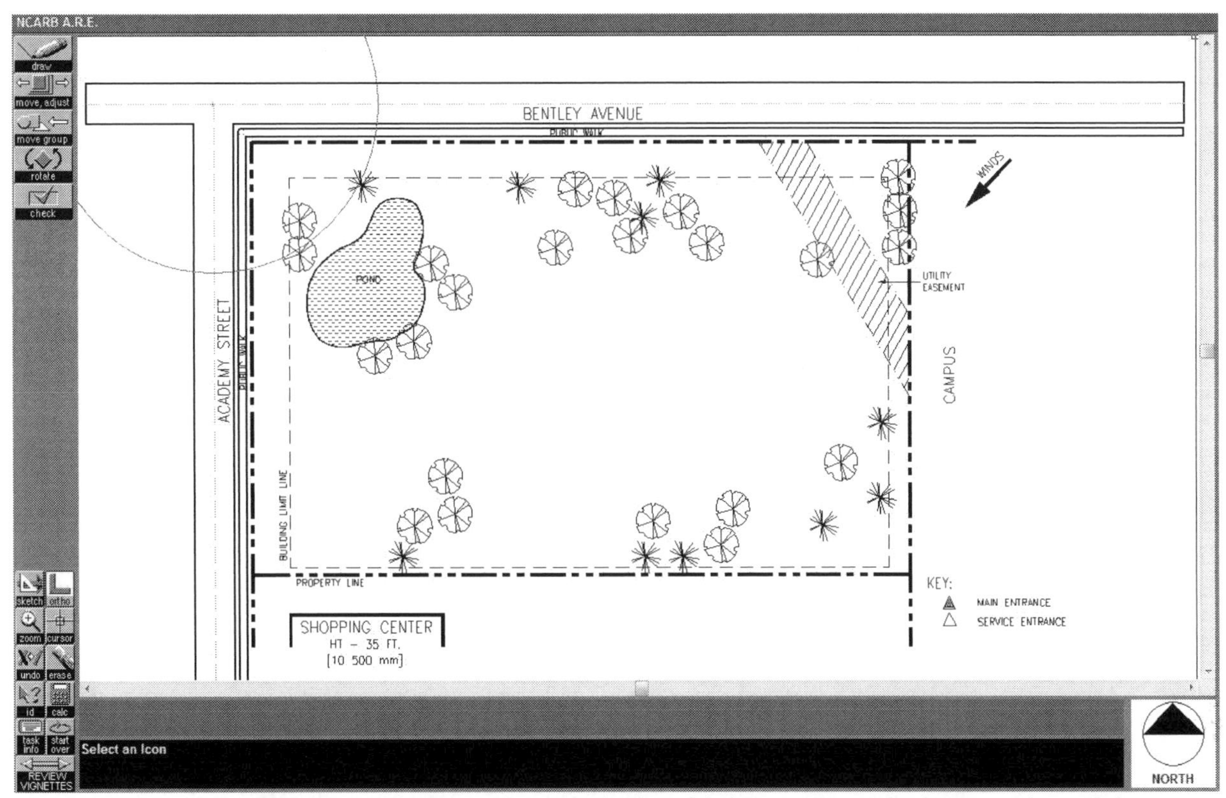

Figure 2.12 Use **Sketch > Circle** to draw a circle with its center at the intersection of the centerlines of the two existing public streets.

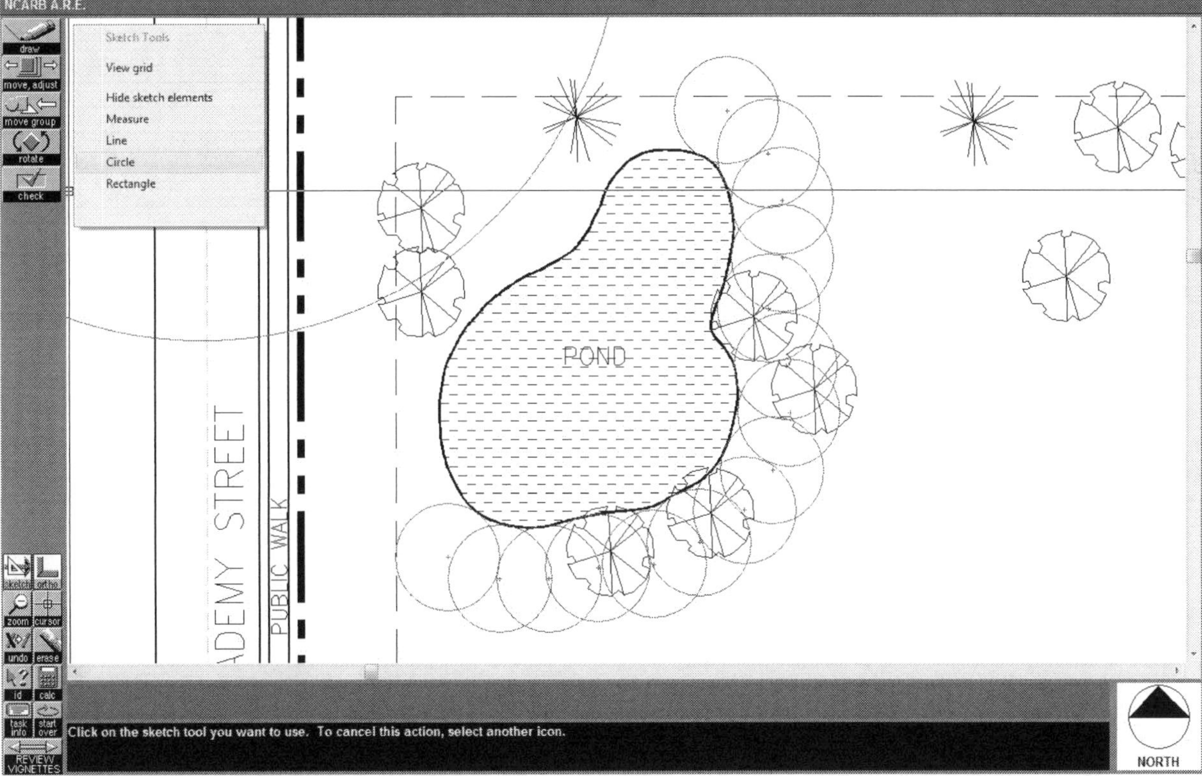

Figure 2.13 Use **Sketch > Circle** to draw a number of circles with 15ft radii defining the 30 ft setback from the pond.

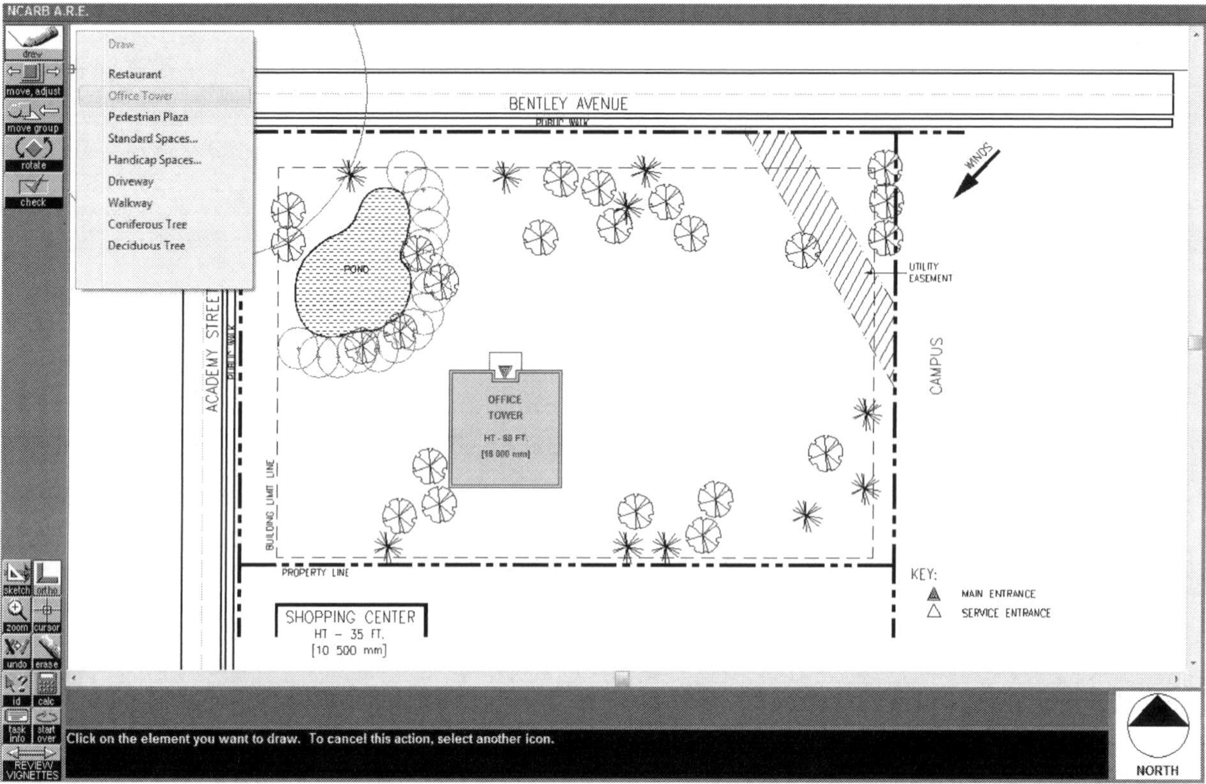

Figure 2.14 Use **Draw > Office Tower** to draw the office tower near the pond with its main entrance facing Bentley Avenue.

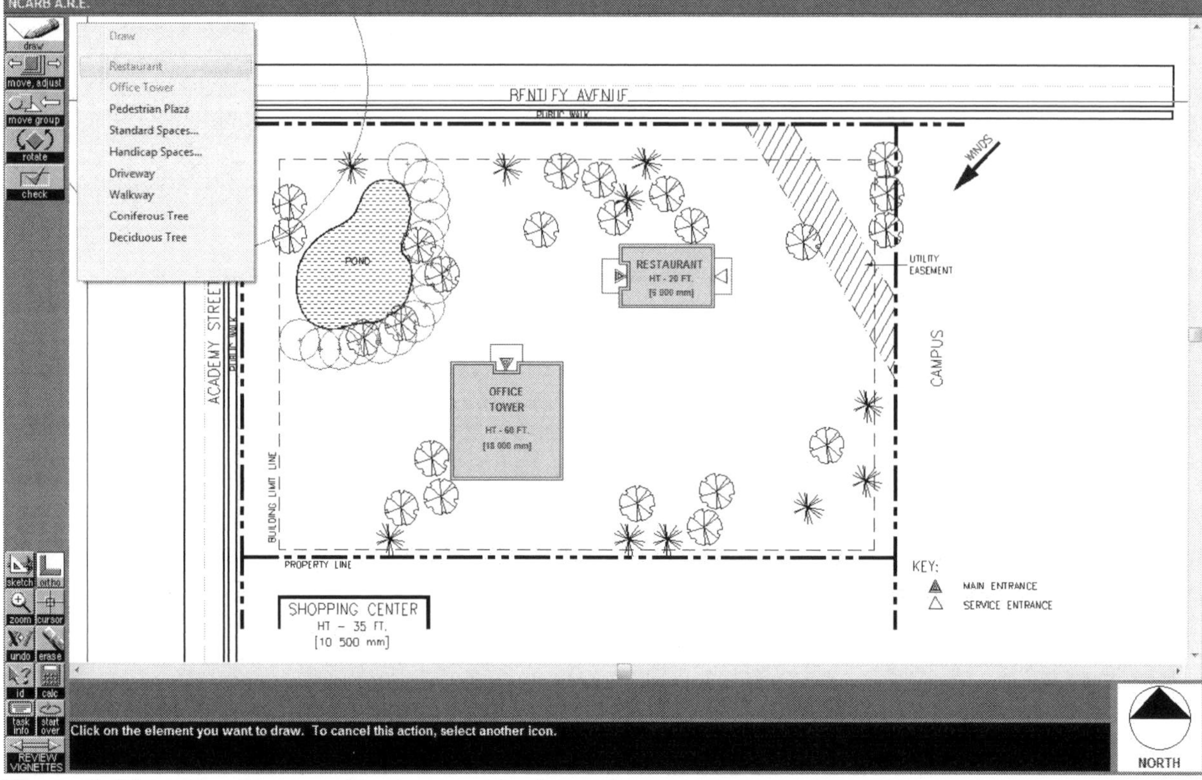

Figure 2.15 Use **Draw > Restaurant** to draw the restaurant.

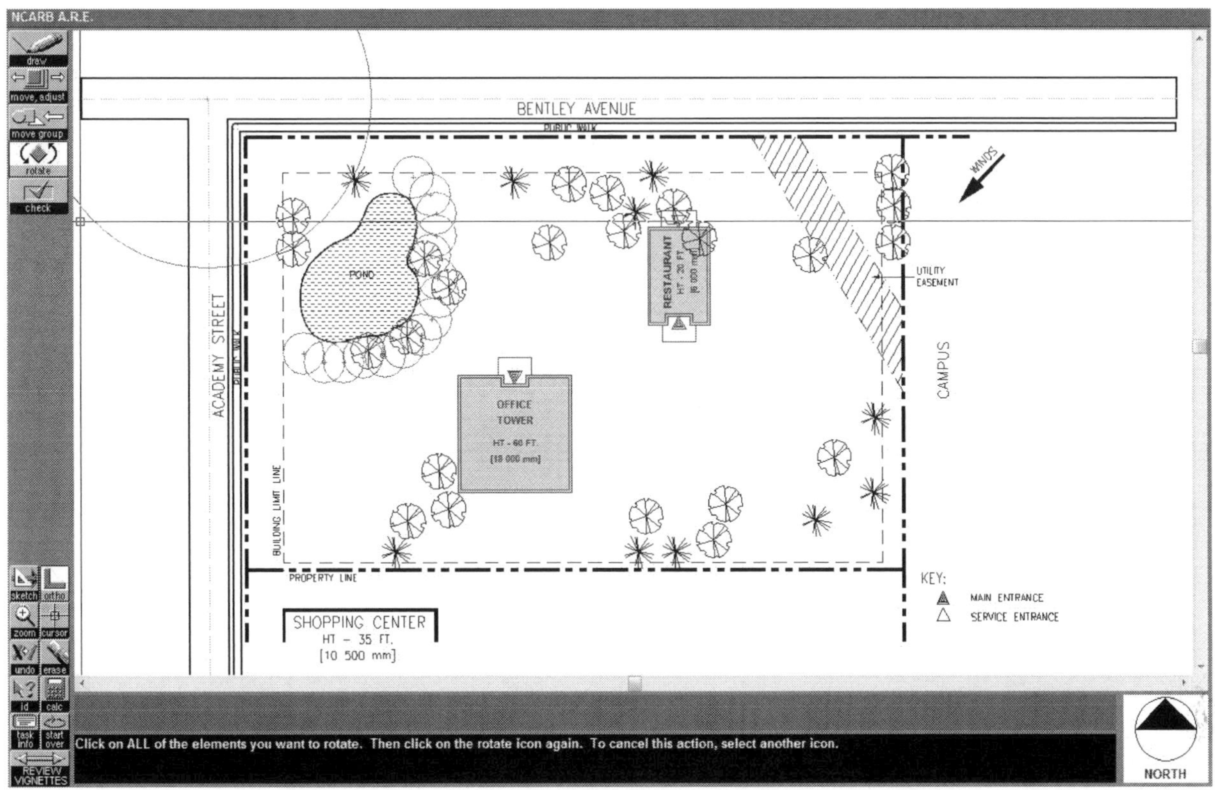

Figure 2.16 Click on **Rotate**, click on the restaurant, and click on **Rotate** again to rotate the restaurant.

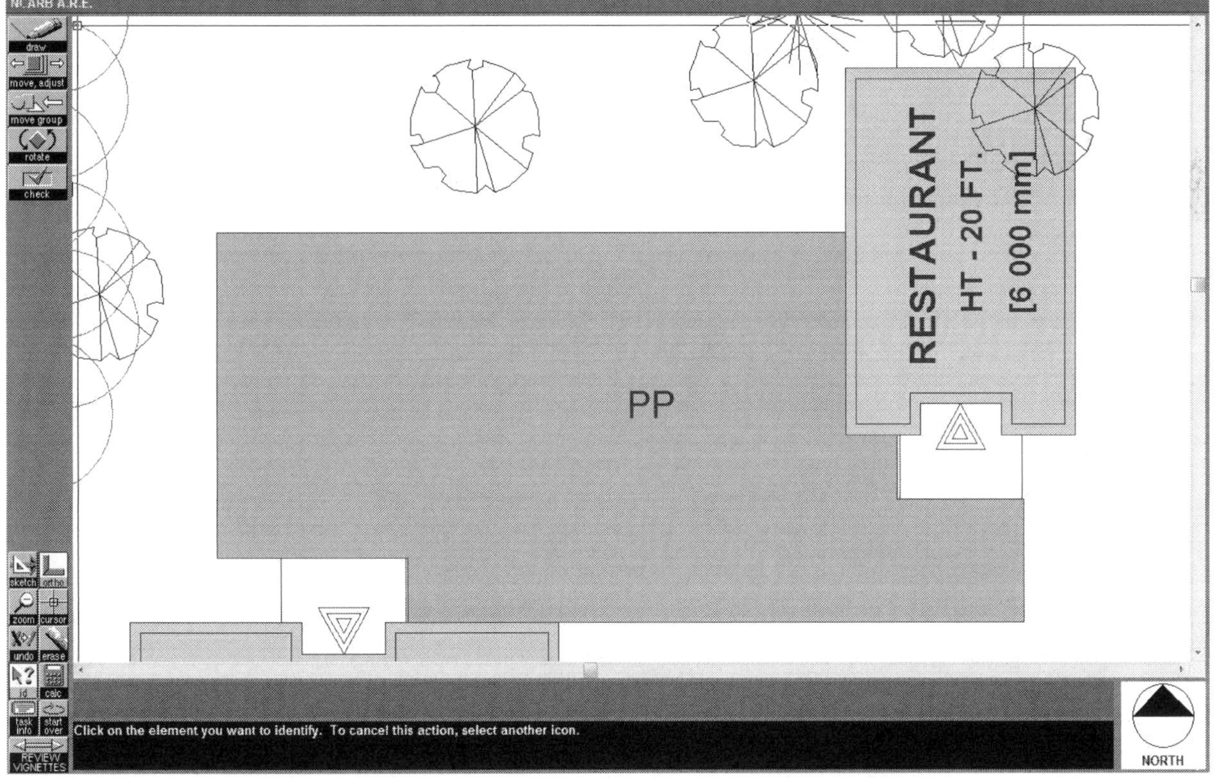

Figure 2.17 Use **Draw > Pedestrian Plaza** to draw the Pedestrian Plaza.

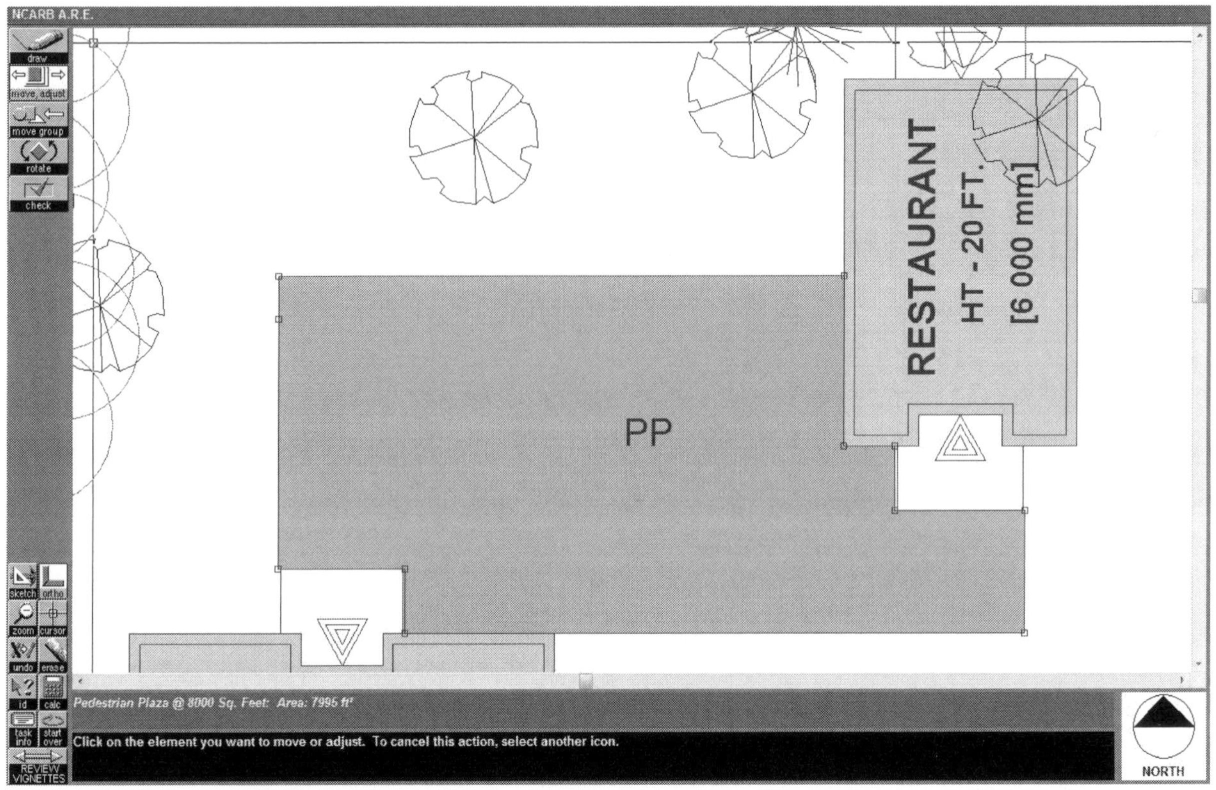

Figure 2.18 Use **Move, Adjust** to adjust the square footage to as close to 8,000 sf, the required square footage, as possible.

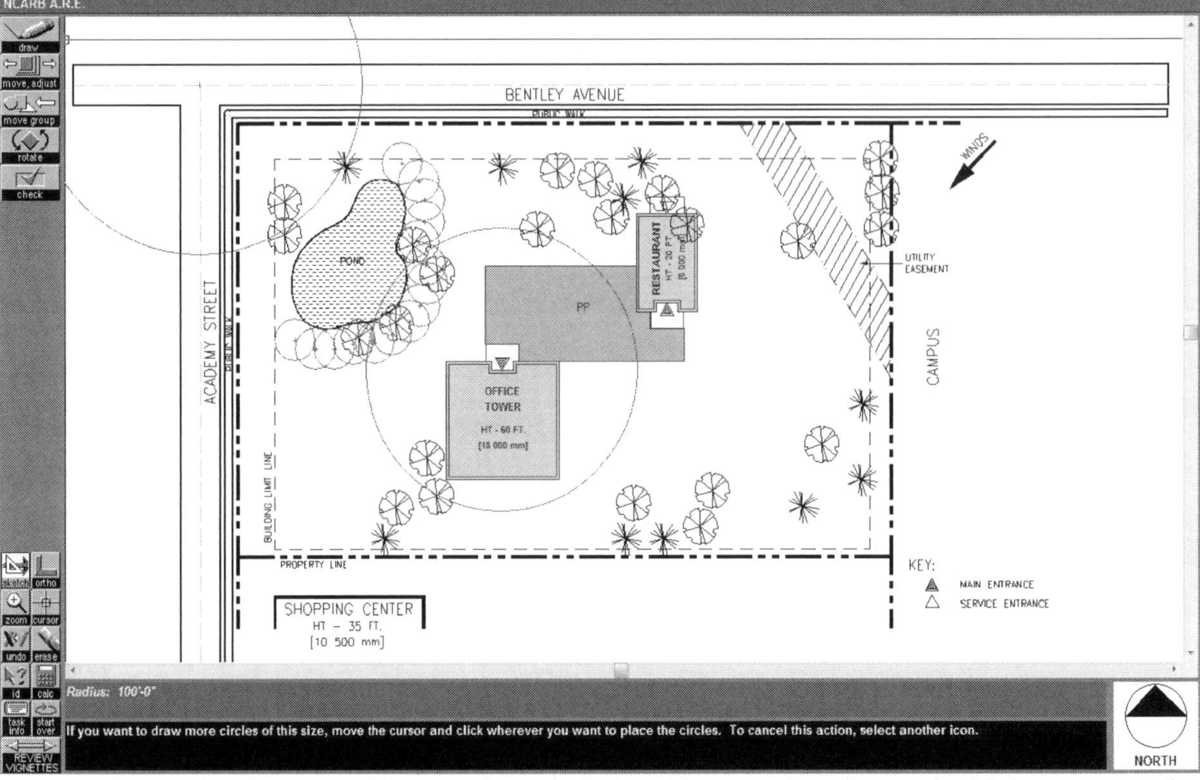

Figure 2.19 Use **Sketch > Circle** to draw a circle with a 100-ft radius and with its center at the main entrance of the Office Tower.

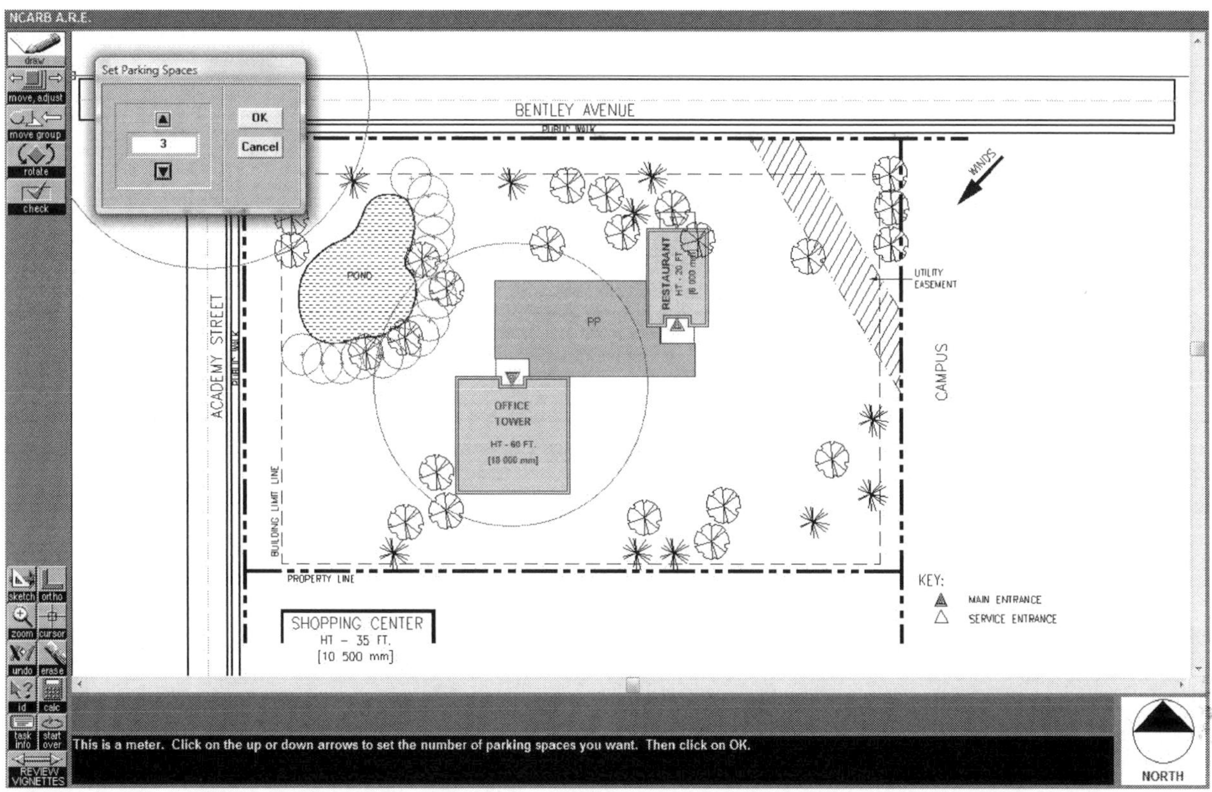

Figure 2.20 Click on the down arrow to set the number of Handicap Parking spaces to 3.

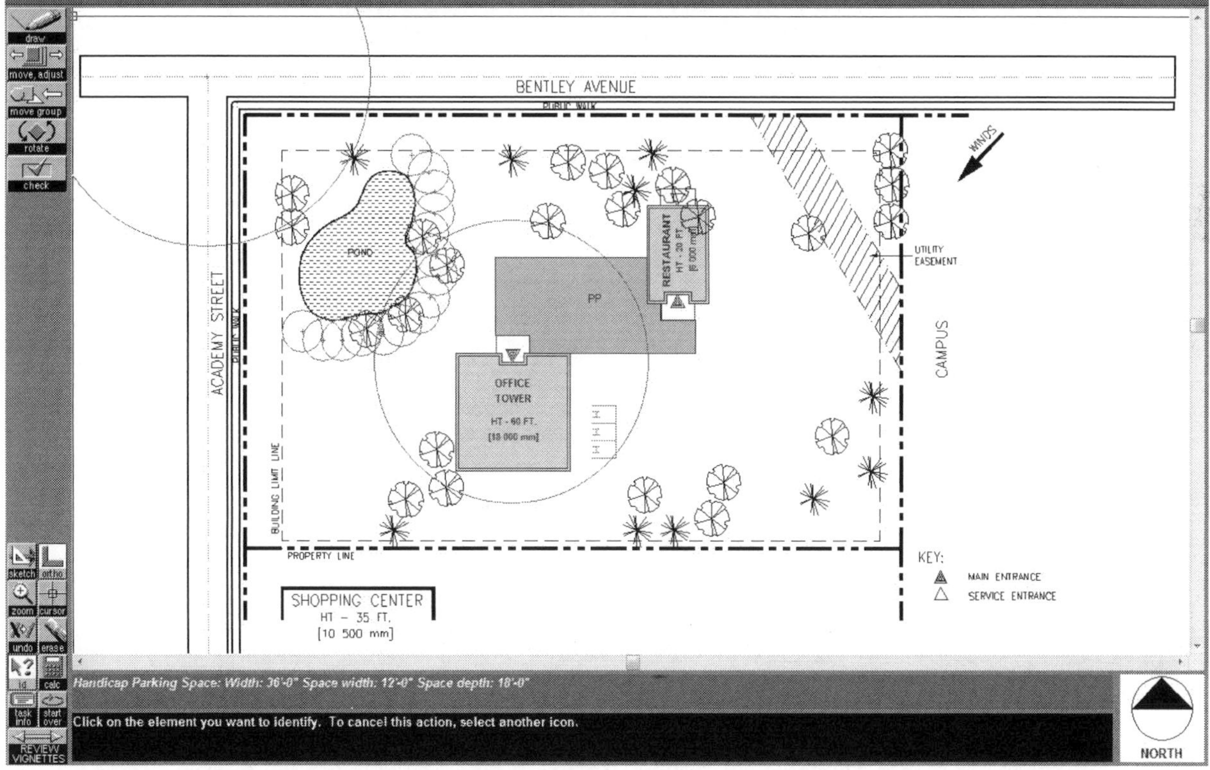

Figure 2.21 Set the over widths of the Handicap Spaces to 36' (12'x3=36'), and the depth of the Handicap Spaces to 18'.

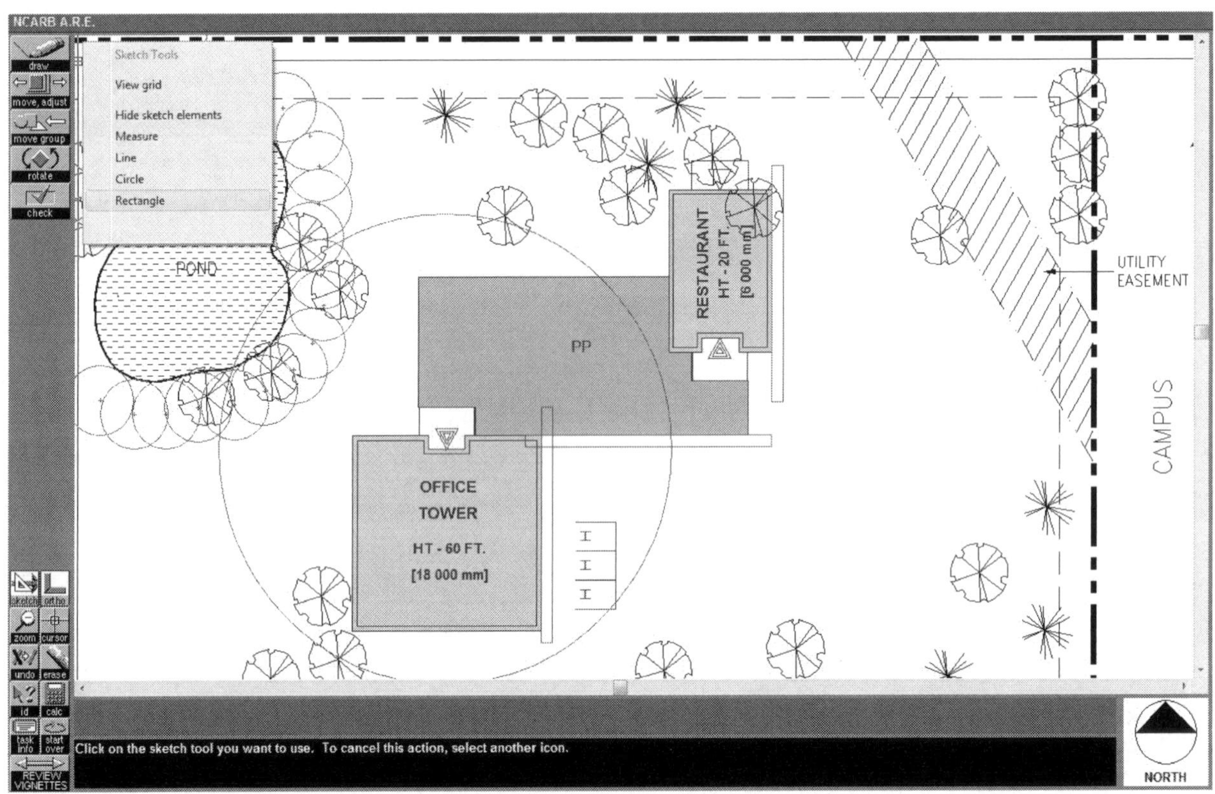

Figure 2.22 Use **Draw > Sketch Rectangle** to draw the 5' wide sketch rectangles to set the 5 -ft clearance for the Office Tower and the Restaurant.

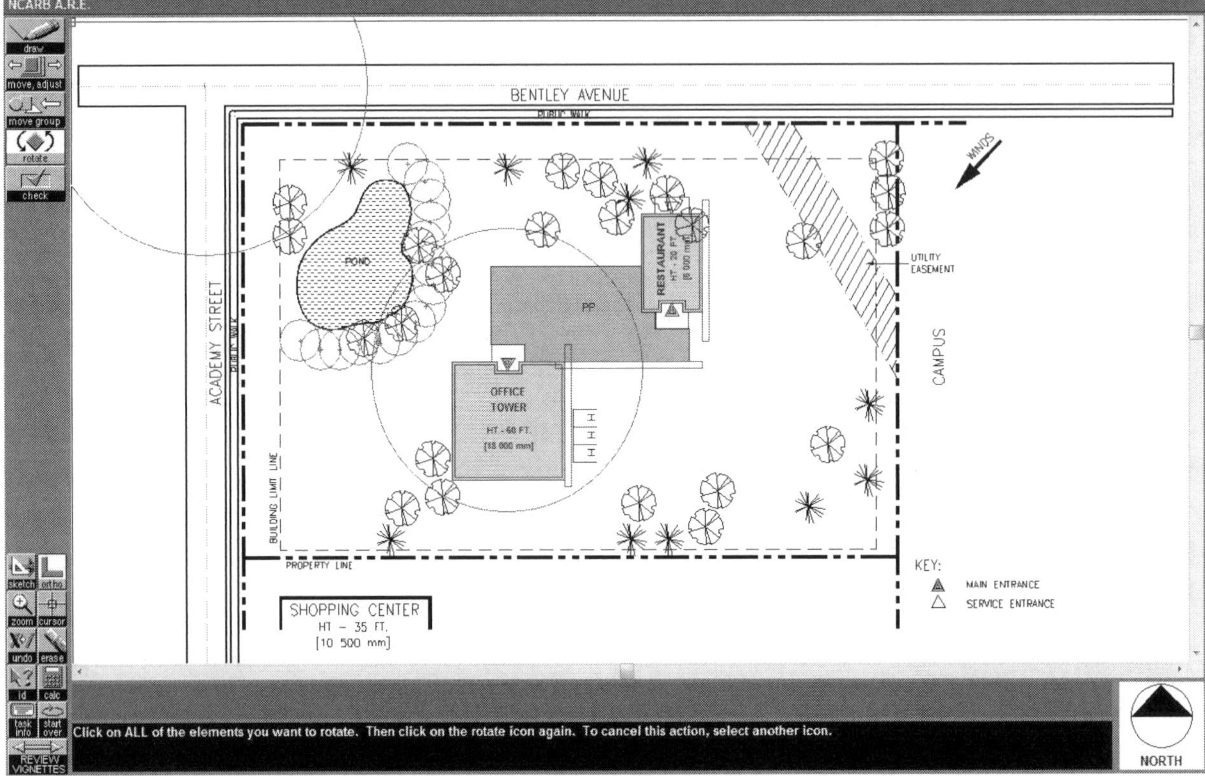

Figure 2.23 Use **Rotate** to rotate the 3 Handicap Spaces. Use **Move, Adjust** to move them closer to the Office Tower.

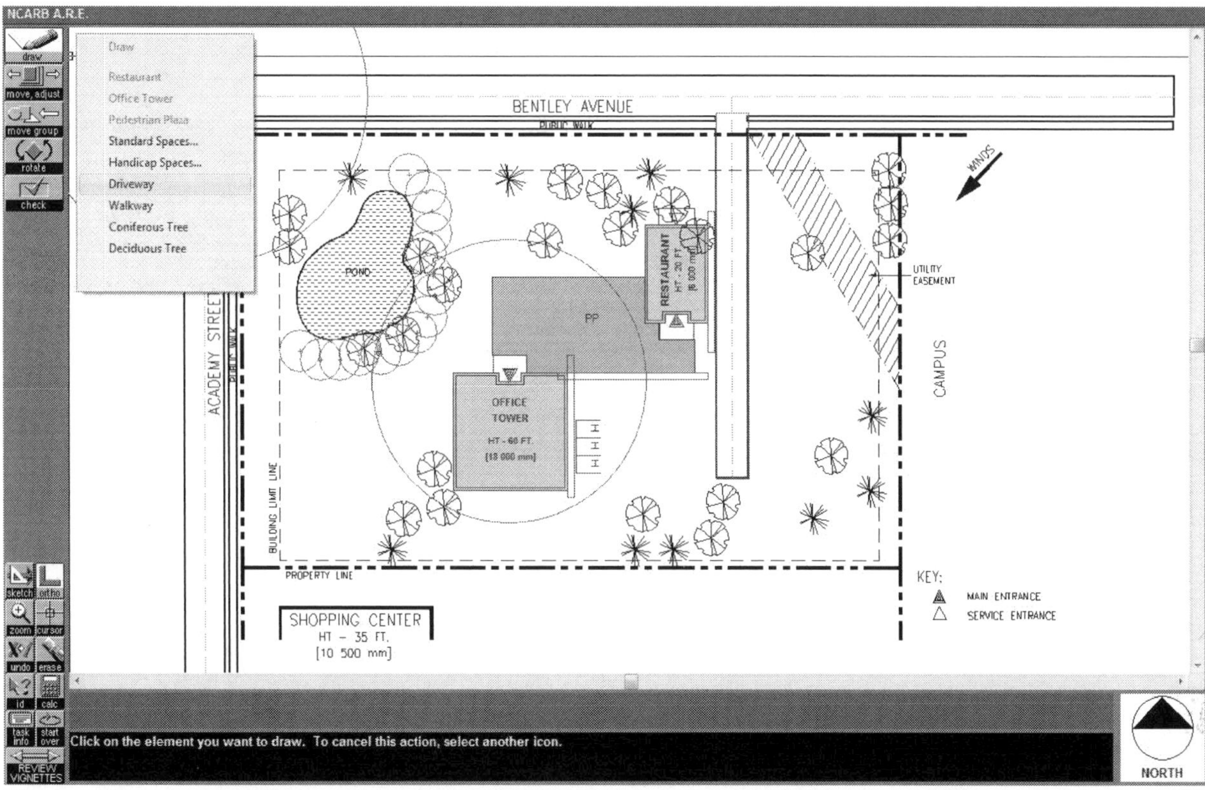

Figure 2.24 Use **Draw > Driveway** to draw a drive from Bentley Avenue, to the left of the utility easement, and at least 5' from the Restaurant.

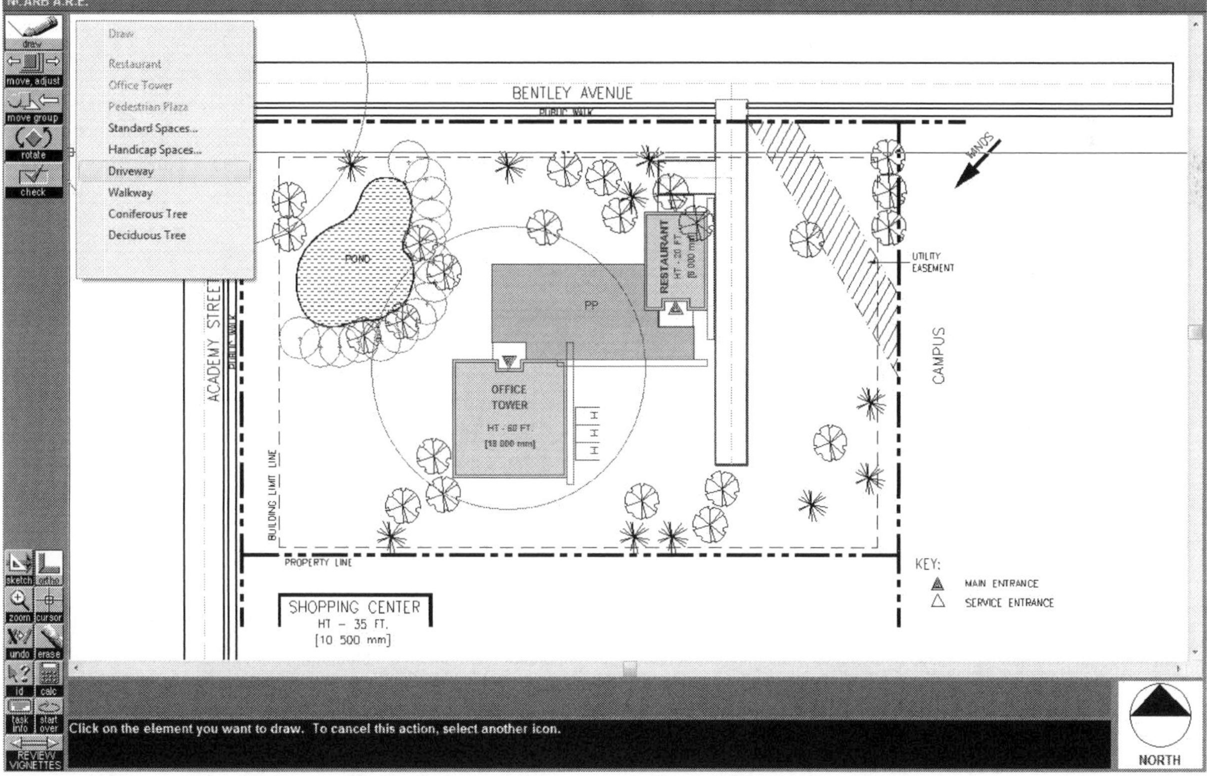

Figure 2.25 Use **Draw > Driveway** to draw a service drive to the service entrance of the Restaurant.

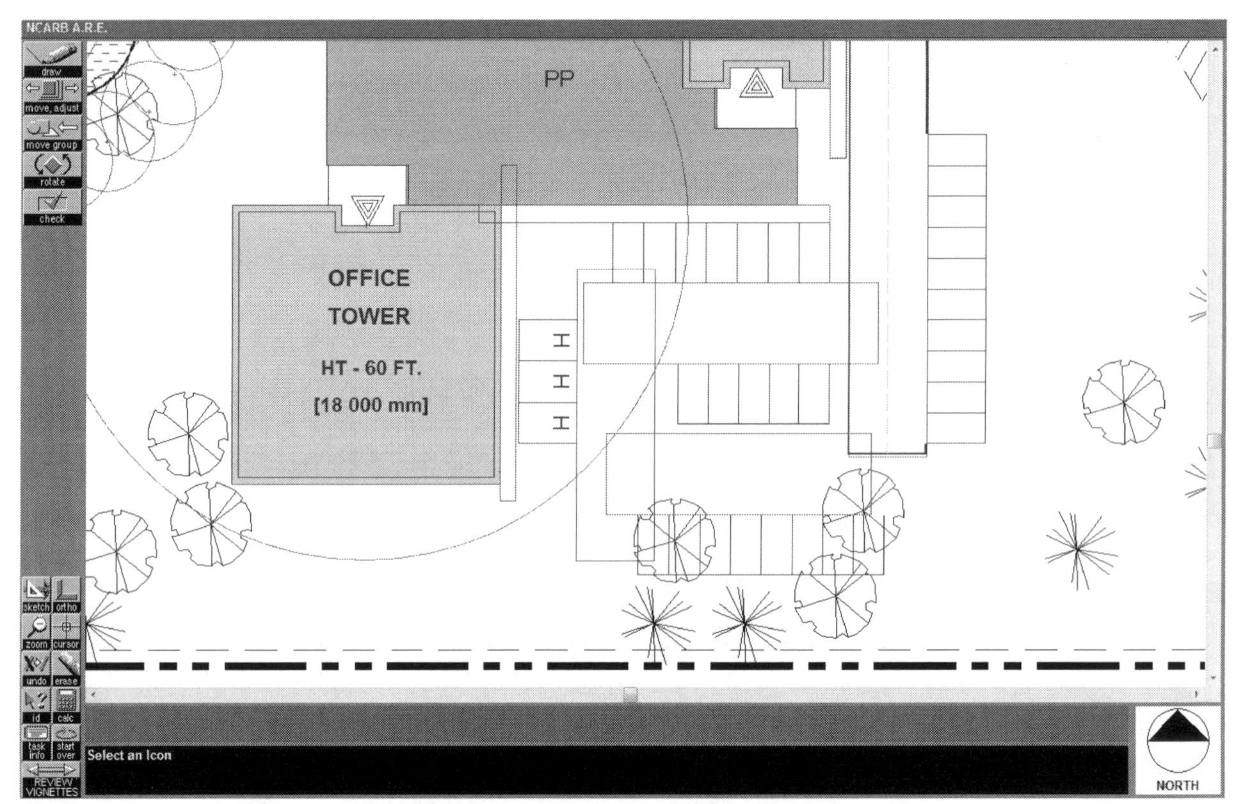

Figure 2.26 Use **Draw > Standard Spaces** to draw 30 standard spaces.

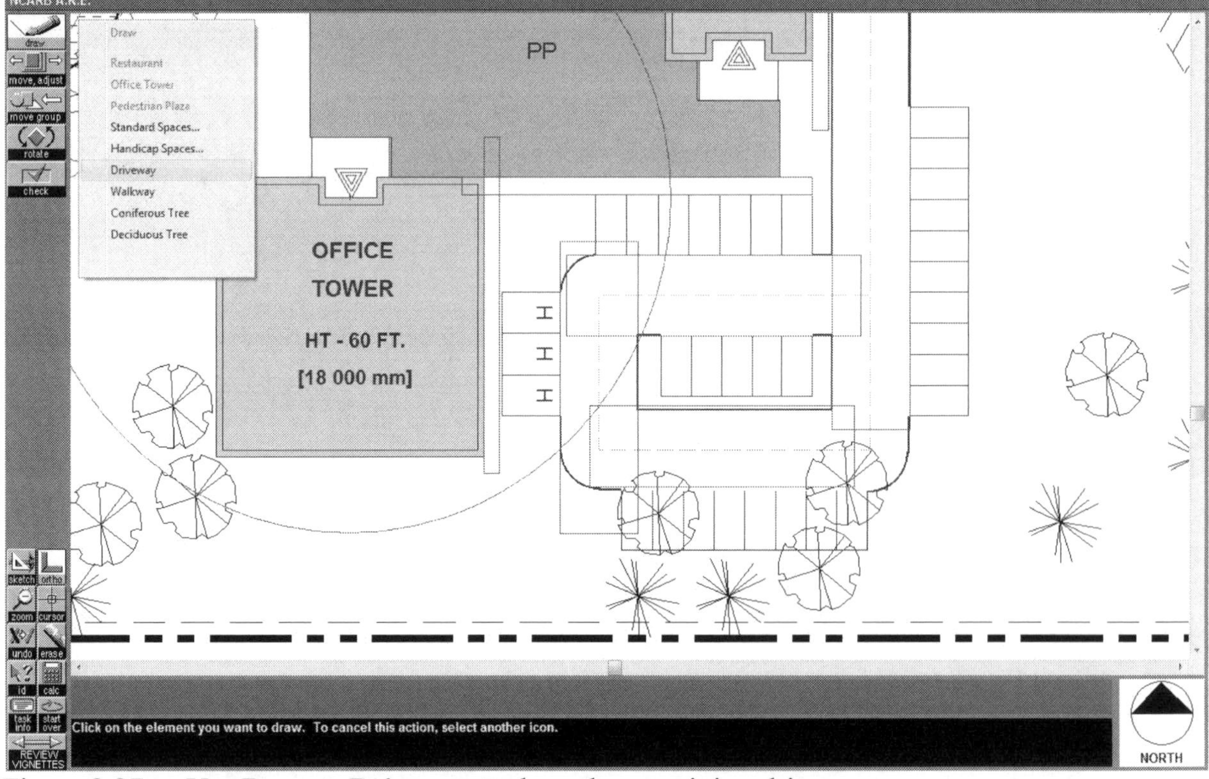

Figure 2.27 Use **Draw > Driveway** to draw the remaining driveways.

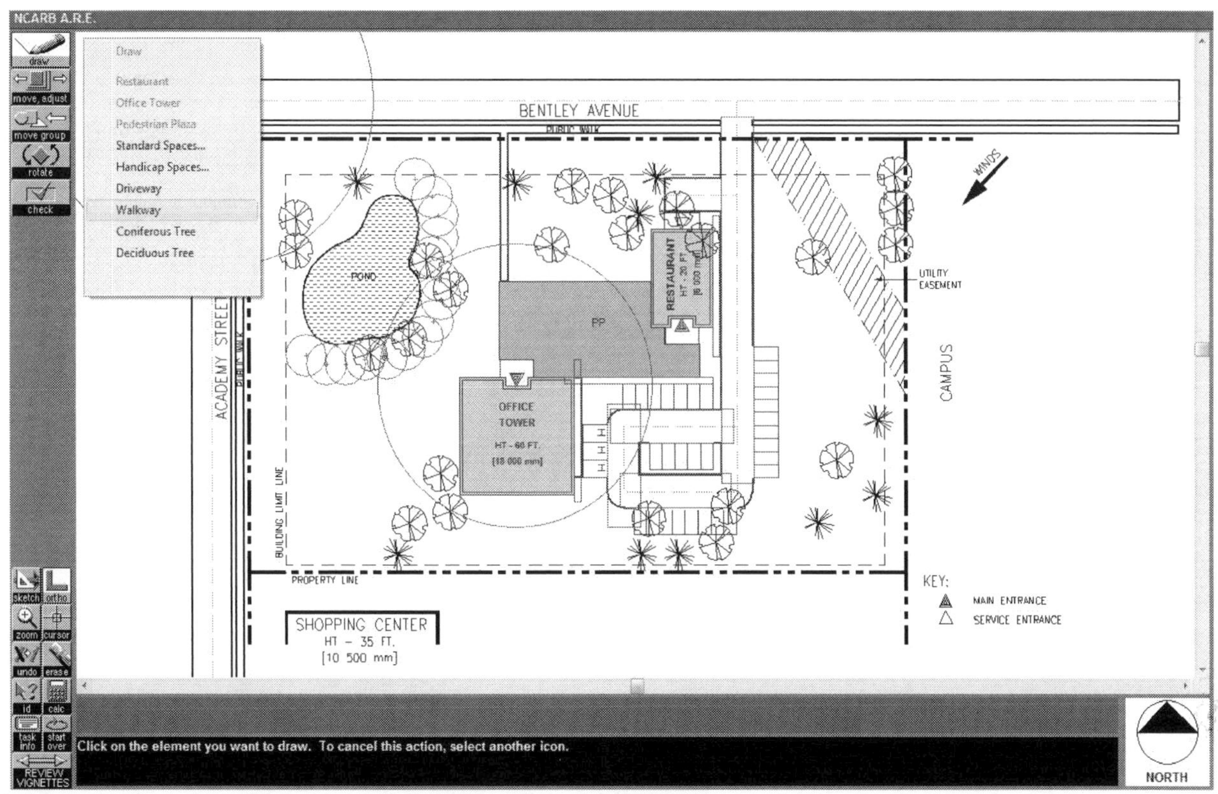

Figure 2.28 Use **Draw > Sidewalk** to draw sidewalks.

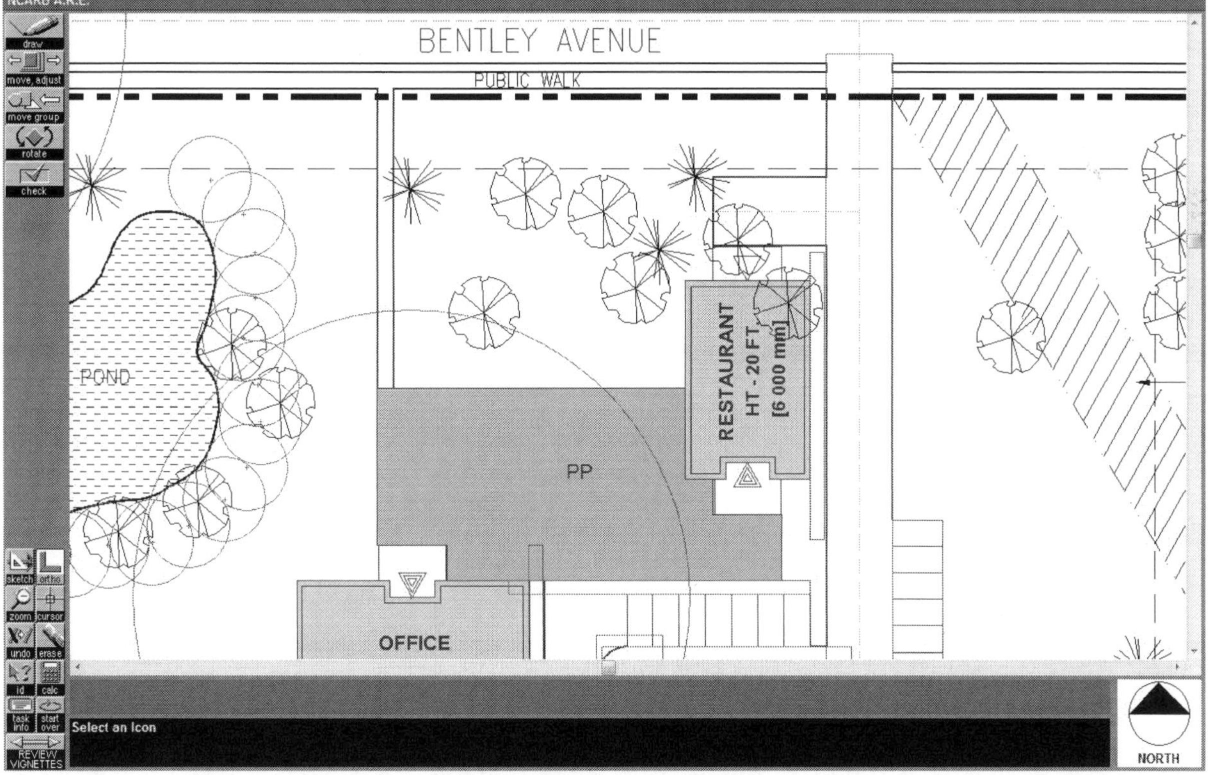

Figure 2.29 The service entrance is NOT facing the Pedestrian Plaza. It is already blocked by the Restaurant itself from the Pedestrian Plaza.

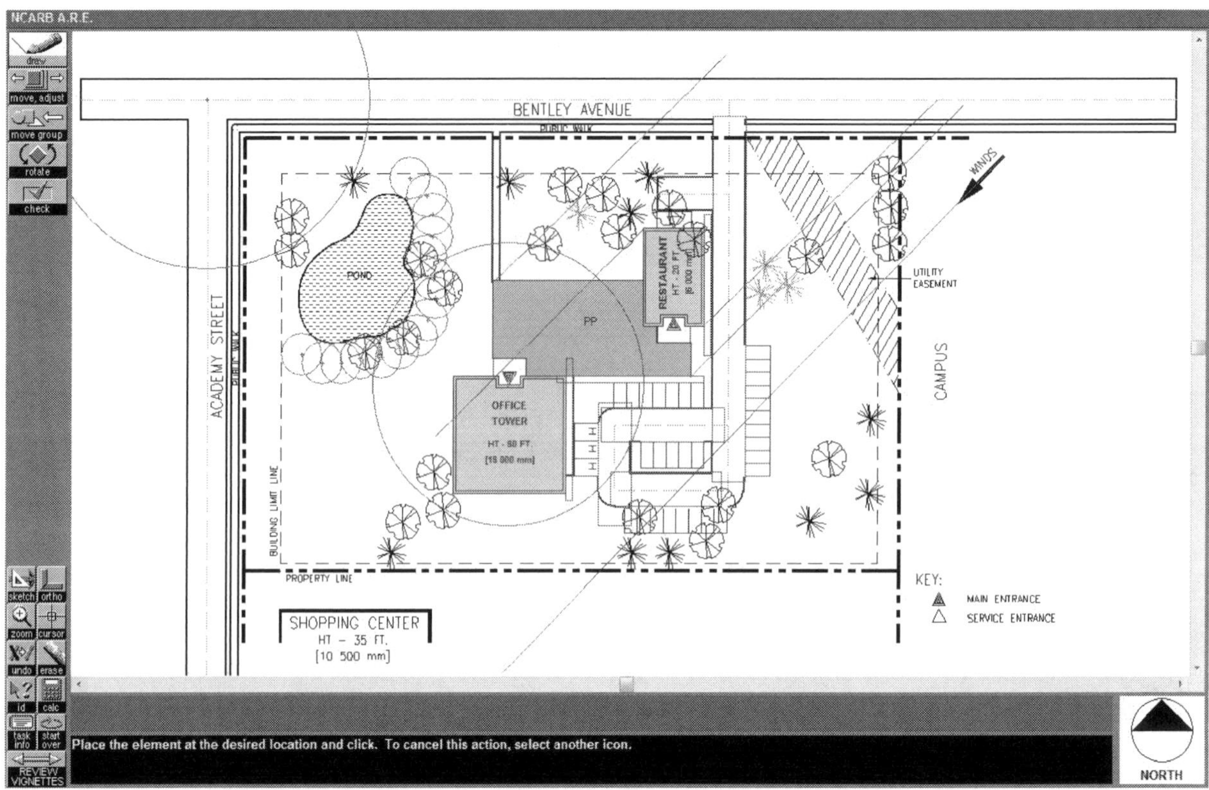

Figure 2.30 Use **Sketch > Line** to draw a few sketch lines parallel to the direction of wind to assist us to locate the coniferous trees.

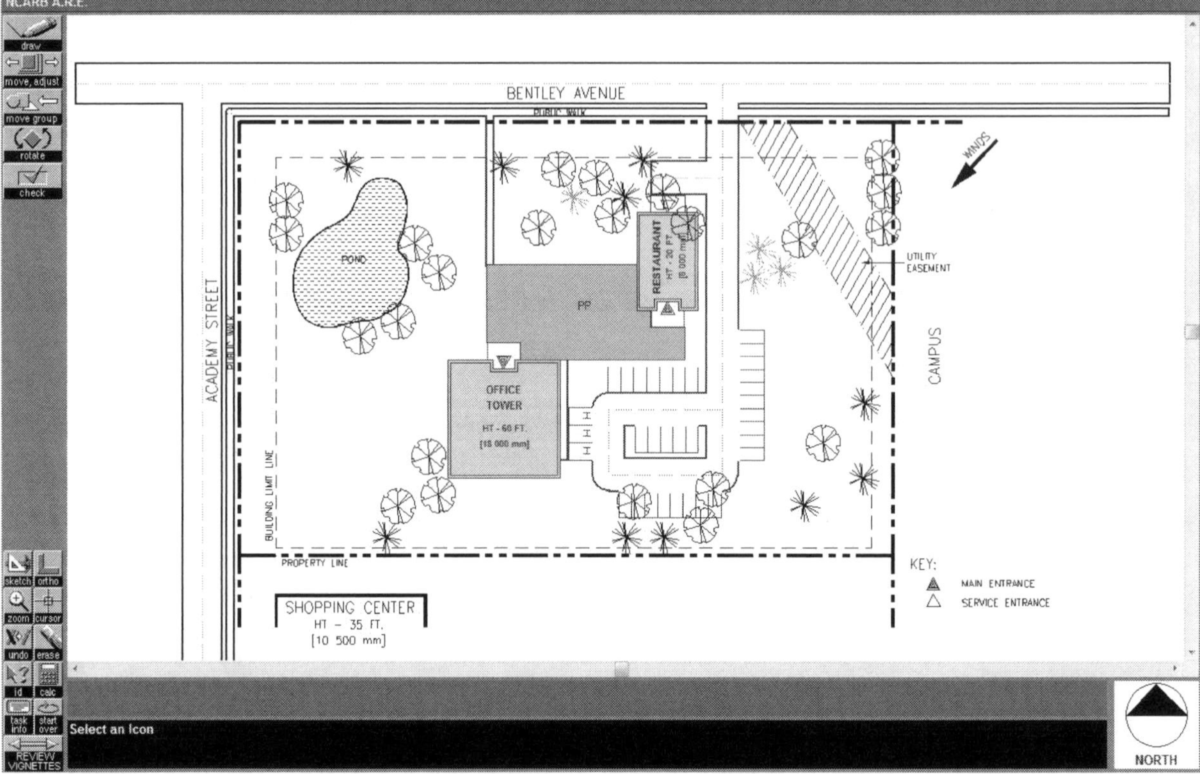

Figure 2.31 We notice the bank of standard parking spaces along the south side of the Pedestrian Plaza is at a wrong orientation.

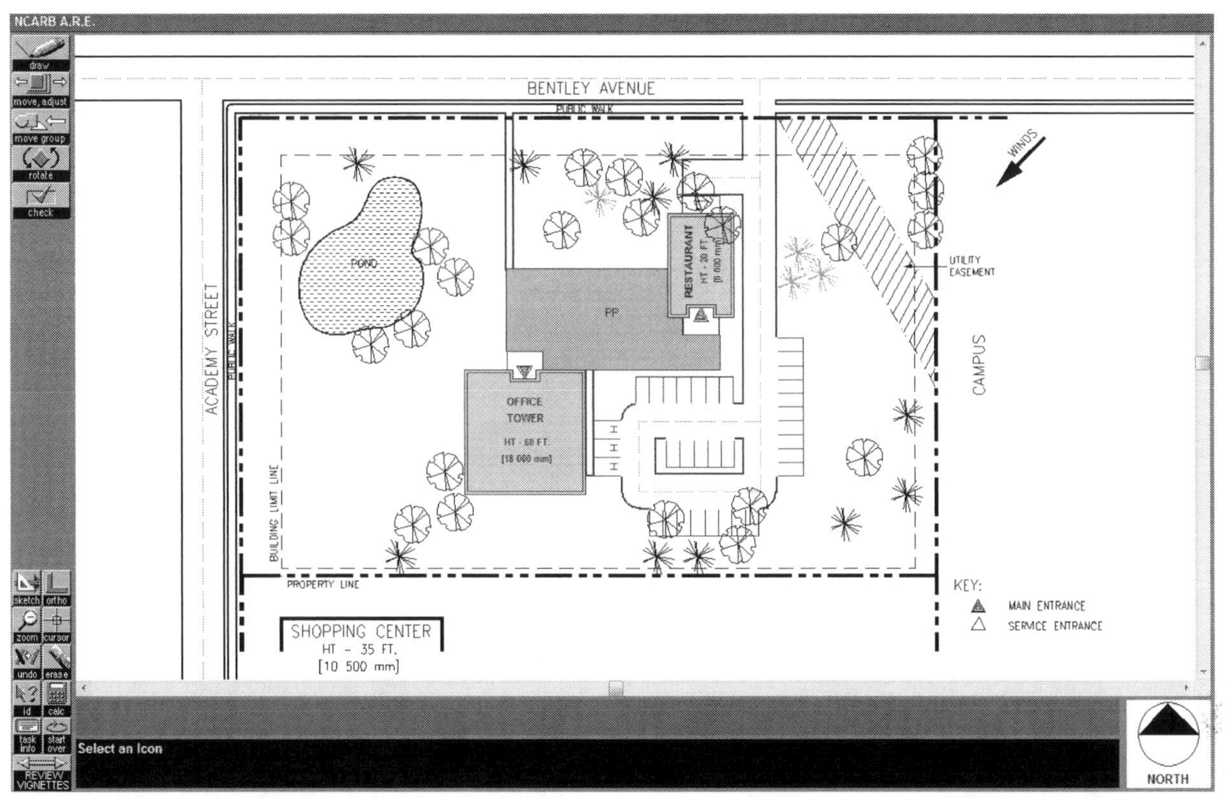

Figure 2.32 Use **Rotate** to rotate the parking spaces 180 degrees, and use **Move Group** to move them to the right location.

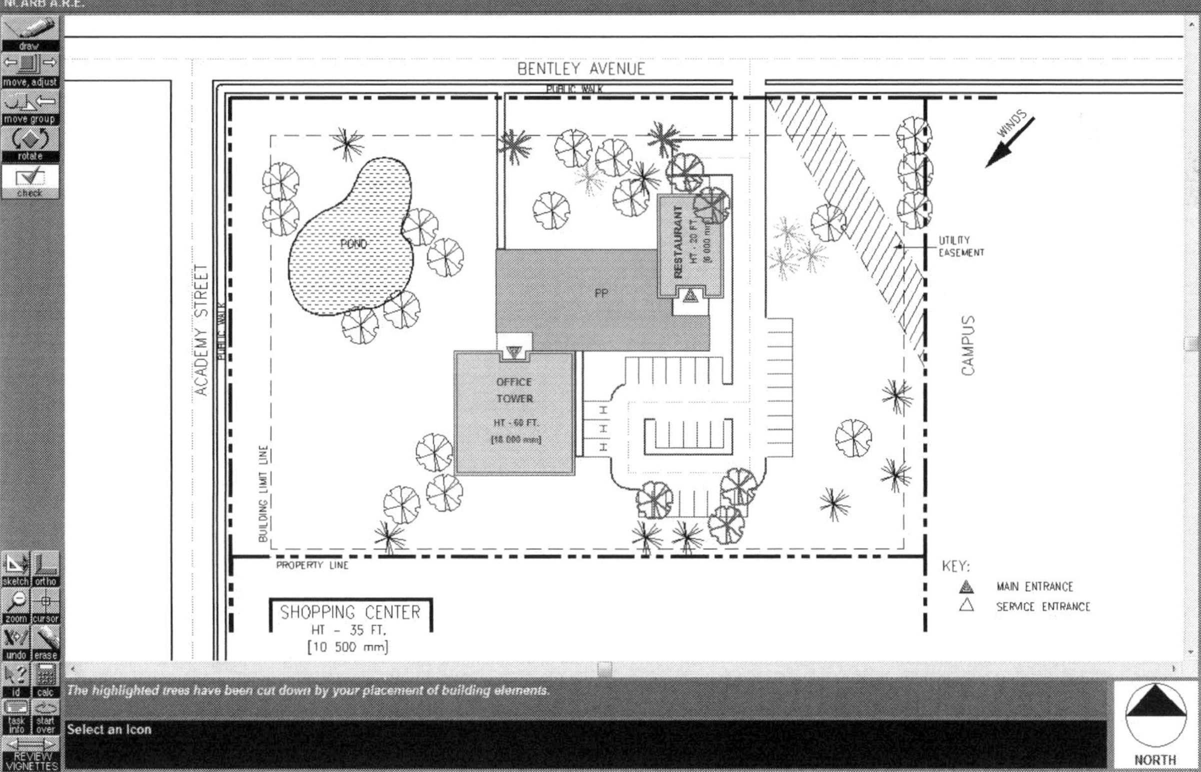

Figure 2.33 Click on **Check**, all the trees that overlap with sidewalk, parking space or other elements will be highlighted.

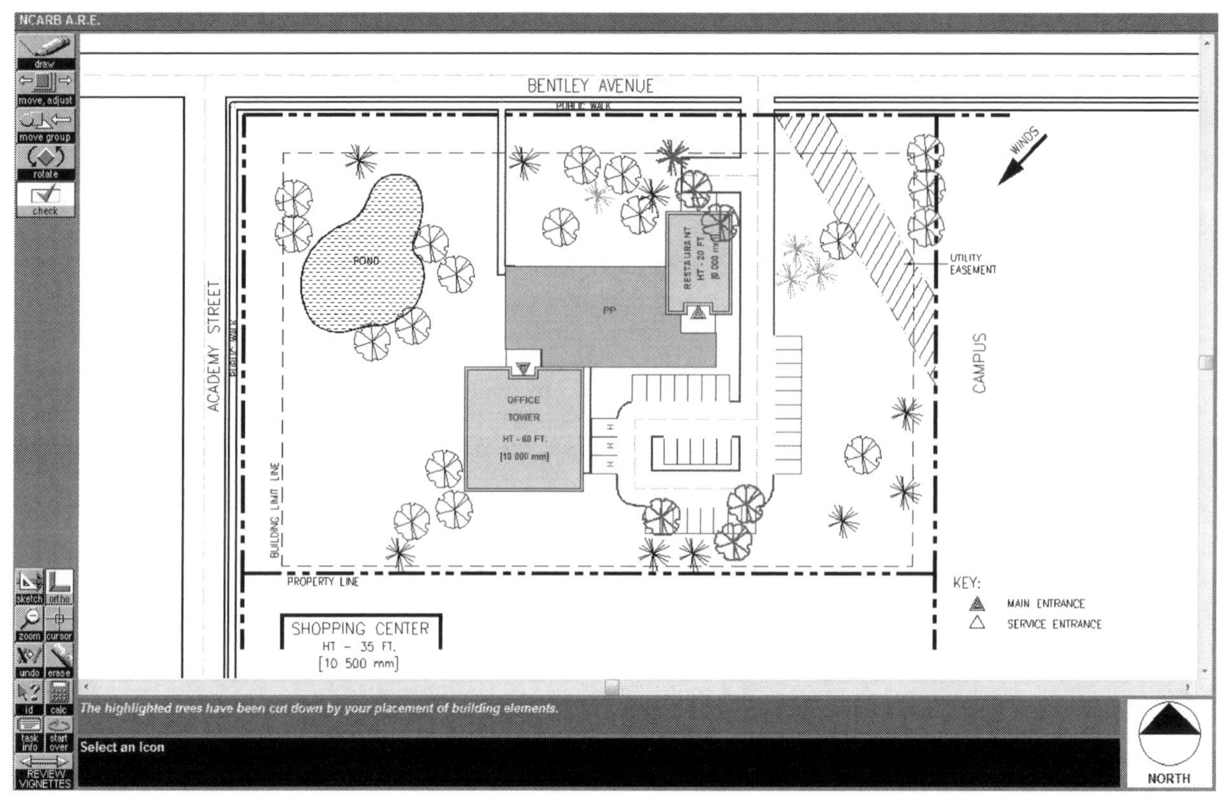

Figure 2.34 If you are NOT comfortable with having 7 highlighted trees, you can adjust the walkway between the Pedestrian Plaza and the public walkway.

6. **Items that you need to pay special attention to**

Several items that you need to pay special attention to:

1) Use **Bubble Diagram** to save time.
2) Use **Sketch** to assist you to locate the elements.
3) A circle with a 15-ft radius has a 30 ft diameter, which equals the required 30 ft setback from the pond.
4) Use *only* **Draw > Handicap Spaces** to draw handicap spaces. If you accidentally use **Draw >Standard Spaces instead**, you need to erase them and redraw them. There is no way to adjust Standard Spaces and change them into handicap spaces.
5) The intersection of the access drive with the street must be perpendicular to the street for at least the first 20 ft of the drive.
7) The key to draw, align or join different sections of the driveway is to align the centerlines of the driveways.
8) Use **Move, Adjust** to adjust the length of the driveways if necessary.
9) Sometimes it may be easier to erase a driveway and then redraw it instead of trying to adjust it.
10) The shortest road segment you can draw is 20'-0".
11) Only coniferous trees can block the view all year long. Only coniferous trees can block the prevailing winter winds.
12) Drive-through circulation is required. Dead-end parking is prohibited. Parking along the service drive is prohibited.

13) No construction or built improvements except for driveways and walkways shall occur *outside* the building limit line.
14) Never place buildings *inside* the easement.
15) Pay attention to the size of the Pedestrian Plaza. Draw the Pedestrian Plaza as close to the required size as possible. In any case, it should NOT be 10% larger or smaller than the required size.
16) The sketch lines parallel to the direction of wind are very valuable tool for you to layout the coniferous trees.

Chapter Three

ARE Mock Exam for Site Planning & Design (SPD) Division

A. Mock Exam: SPD Multiple-Choice (MC) Section

1. The purpose of a percolation test is to determine the
   a. absorption rate
   b. soil alkalinity
   c. soil density
   d. gravel sizes
   e. evaporation rate
   f. absorption rate of soils

2. Which of the following loads defines the force of rainwater exerted on roof structures?
   a. Hydrostatic
   b. Dynamic
   c. Water
   d. Wedge
   e. Live

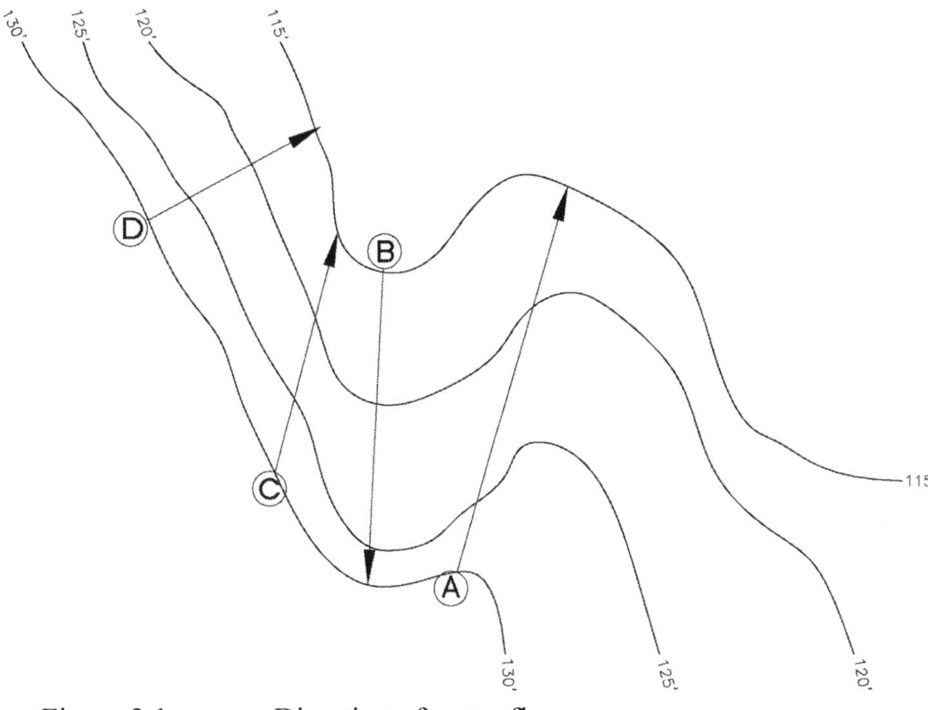

Figure 3.1 Direction of water flow

3. Refer to Figure 3.1, which of the following is the direction of water flow?
   a. A
   b. B
   c. C
   d. D

4. An architect needs to set the spot elevations at the perimeter of a one-story supermarket building. If the finish floor elevation is 112.00', what is a proper spot elevation at the front entrance if the front door has a threshold?
   a. 111.75'
   b. 111.96'
   c. 112.00'
   d. 112.25'

5. For the same one-story supermarket building in previous question, what is a proper spot elevation at the front entrance if the front door has no threshold?
   a. 111.75'
   b. 111.98'
   c. 112.00'
   d. 112.25'

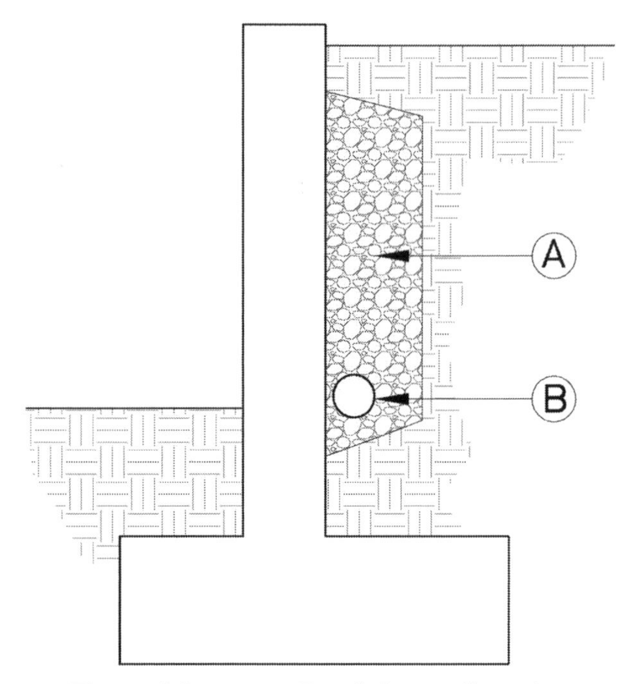

Figure 3.2 Retaining wall section

6. Which of the following is A as shown on Figure 3.2?
   a. Sand
   b. Soil
   c. Sand and gravels
   d. Gravel

7. Which of the following is B as shown on Figure 3.2?
   a. Sewer line
   b. Storm drain
   c. Perforated pipe
   d. Water line

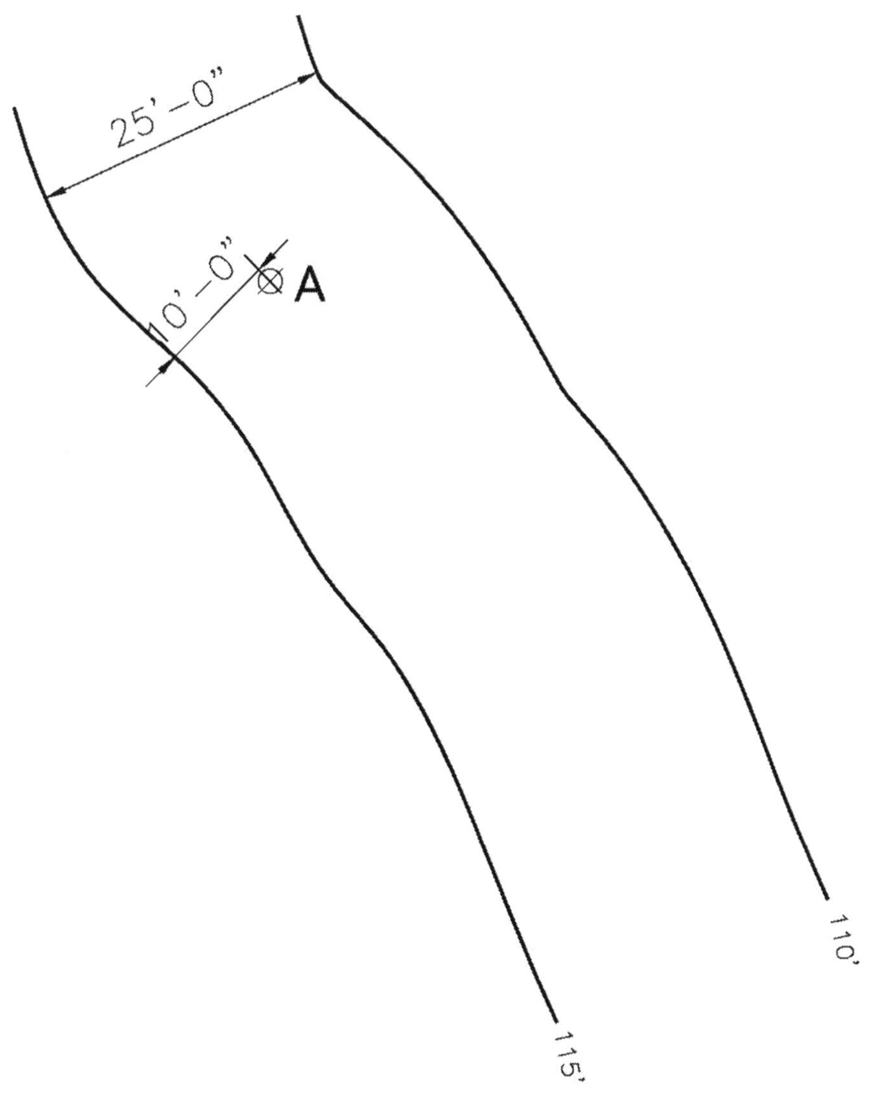

Figure 3.3 Spot elevations

8. Assuming a constant slope between the two contour lines shown, what is the elevation of point A?
   a. 112'
   b. 112.5'
   c. 113'
   d. 113.5'
   e. 114'

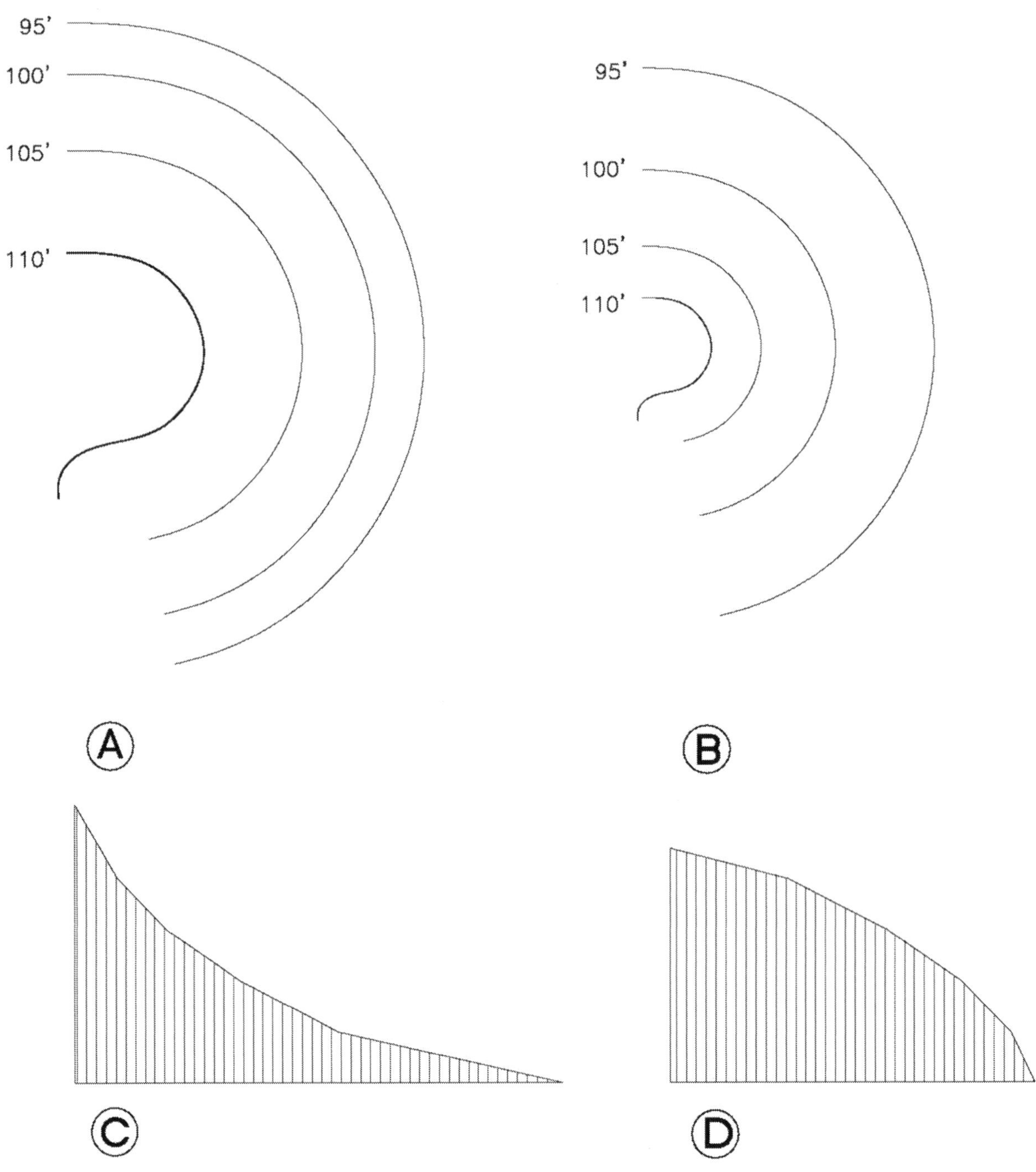

Figure 3.4 Types of slopes

9. Refer to Figure 3.4, which of the following shows a convex slope? **Check the two that apply.**
   a. A
   b. B
   c. C
   d. D

10. Refer to Figure 3.4, which of the following shows a concave slope? **Check the two that apply.**
    a. A
    b. B
    c. C
    d. D

11. Which of the following is true? **Check the two that apply.**
    a. In Southern Hemisphere, the solar altitude angle is lowest in December and highest in June.
    b. In Southern Hemisphere, the solar altitude angle is highest in December and lowest in June.
    c. In Northern Hemisphere, the solar altitude angle is lowest in December and highest in June.
    d. In Northern Hemisphere, the solar altitude angle is highest in December and lowest in June.

12. Which of the following is the most energy efficient orientation of a rectangular building in Los Angeles?
    a. Locate the building with the long direction oriented along the north-south axis and facing slightly to the east.
    b. Locate the building with the long direction oriented along the north-south axis and facing slightly to the west.
    c. Locate the building with the long direction oriented along the east-west axis and facing slightly to the east.
    d. Locate the building with the long direction oriented along the east-west axis and facing slightly to the west.

13. An architect is working on a room addition project. The existing single family home has 1,600 s.f. and 2 existing toilets. The owner wants to add 1,000 s.f. of new area, including 3 new bedrooms and 2 new toilets. The existing 3" interior sewer line connects to the 4" main sewer line at about 6' deep and 3' in front of the house. Which of the following is correct? **Check the two that apply.**
    a. The architect can connect the new toilets to the existing 4" sewer line.
    b. The architect can connect the new toilets to the existing 3" sewer line.
    c. It is better to oversize a sewer line when designing buildings.
    d. The architect can ask his plumbing engineer for advice.

14. If the landing depth at the top of a curb ramp is less than 48 inches, then the slope of the flared side shall not exceed________
    a. 1:20
    b. 1:12
    c. 1:10
    d. 1:8

15. Per *International Building Code* (IBC), Section 1010.6, the minimum width of a means of egress ramp between handrails, if provided, or other permissible projections shall be ______.
    a. 32"
    b. 36"
    c. 48"
    d. 60"

16. Which of the following is unlikely to show up on an ALTA survey plan?
    a. Metes and bounds
    b. Fire hydrants
    c. Trees
    d. Footings
    e. Retaining walls

17. Which of the following is the correct order of arranging the units used in the Public Land Survey System (PLSS) in the US, from large to small?
    a. Check, section, township
    b. Section, township, check
    c. Check, township, section
    d. Section, check, township

18. A developer wants to build an elementary school in an area zoned for residential use. Which of the following applications should she submit to the city?
    a. Conditional Use Permit
    b. Non-conforming Use Permit
    c. Incentive Zoning Permit
    d. Ordinance Variance Permit

19. Which of the following statements are correct? **Check the two that apply.**
    a. In a cold climate, an architect should use materials with low albedo and low conductivity for ground surfaces.
    b. In a cold climate, an architect should use materials with low albedo and high conductivity for ground surfaces.
    c. In a tropical climate, an architect should use materials with low albedo and low conductivity for ground surfaces.
    d. In a tropical climate, an architect should use materials with high albedo and high conductivity for ground surfaces.

20. Per the U.S. Environmental Protection Agency (EPA), using a ________________-based approach to wetland protection ensures that the whole system, including land, air, and water resources, is protected.

21. Per EPA, which of the following are general categories of wetlands found in the U.S.? **Check the four that apply.**
    a. Bogs
    b. Fens
    c. Lakes
    d. Marshes
    e. Reservoirs
    f. Swamps

22. Per EPA, which of the following is the leading cause of species extinction?
    a. Human activities
    b. Pollution
    c. Habitat degradation
    d. Excessive hunting

23. What is SWPPP?
    a. Solid Waste Pollution Prevention Plan
    b. Sewer Waste Pollution Prevention Plan
    c. Stormwater Pollution Prevention Plan
    d. Soiled Water Pollution Prevention Plan

24. Which of the following is not the purpose of SWPPP?
    a. To prevent stormwater contamination
    b. To prevent airborne dust from getting into rivers
    c. To control sedimentation and erosion
    d. To comply with the requirements of the Clean Water Act
    e. To comply with the requirements of CGP
    f. None of the above

25. What is NPDES?
    a. National Pollutant Discharge Elimination System
    b. National Pollutant Discharge and Exchange System
    c. Natural Pollutant Discharge Elimination System
    d. Natural Pollutant Discharge and Exchange System

26. Who is typically in charge of preparing a Best Management Practices (BMP) plan?
    a. Owner
    b. Architect
    c. Civil Engineer
    d. Contractor

27. Which of the following are structural BMPs? **Check the four that apply.**
    a. Fences
    b. Maintaining equipment
    c. Sedimentation ponds
    d. Erosion control blankets
    e. Temporary or permanent seeding
    f. Training site staff on erosion and sediment control practices

28. Which of the following are incorrect? **Check the two that apply.**
    a. It is usually easier and less expensive to prevent erosion than it is to control sedimentation.
    b. It is usually easier and less expensive to control sedimentation than it is to prevent erosion.
    c. A good SWPPP will use both kinds of BMPs in combination for the best results.
    d. A good SWPPP will use one kind of BMPs for the best results.

29. Which of the following is typically not included in a SWPPP?
    a. Site plans
    b. Maintenance/inspection procedures
    c. Description of controls to reduce pollutants
    d. Plumbing plans

30. Which of the following are objectives of SWPPP? **Check the two that apply.**
    a. Reduce impervious surfaces and promote infiltration
    b. Increase impervious surfaces and promote infiltration
    c. Protect surface waters from erosion and sediment
    d. Protect groundwater from erosion and sediment

31. What does fingerprinting your site mean?
    a. Obtaining the fingerprint of everyone at your site
    b. Obtaining the fingerprint of every visitor to your site
    c. Identifying the trees in your site
    d. Inventorying a site's natural features

32. Which of the following is true regarding SWPPP? **Check the two that apply.**
    a. Silt fencing is more effective than inlet protection.
    b. Silt fencing is less effective than inlet protection.
    c. Hydromulching is more effective than inlet protection.
    d. Hydromulching is less effective than inlet protection.

33. Who must sign the SWPPPs and inspection reports?
    a. Owner
    b. Architect
    c. Civil Engineer
    d. Concrete contractor
    e. Construction Site Operator

34. Which of the following is true regarding Runoff Coefficient? **Check the two that apply.**
    a. Downtown areas tend to have a higher Runoff Coefficient than residential areas.
    b. Downtown areas tend to have a lower Runoff Coefficient than residential areas.
    c. Drives and walks tend to have a higher Runoff Coefficient than lawns.
    d. Drives and walks tend to have a lower Runoff Coefficient than lawns.

35. Which of the following is not a groundwater remediation technology?
    a. Bioventing
    b. Biocontrol
    c. Pump and treat
    d. Air sparging

36. Which of the following is the correct order of senses in conveying information about a site, arranged from most important to least important?
    a. Sight, touch, hearing, smell, taste
    b. Sight, hearing, touch, smell, taste
    c. Sight, hearing, smell, touch, taste
    d. Sight, touch, smell, hearing, taste

37. Which of the following is generally NOT given in a soils report? **Check the two that apply.**
    a. Recommended types of foundations
    b. Slump test results
    c. Structural calculations
    d. Soils bearing capacity
    e. Type of concrete to use

38. Which of the following is likely to cause differential settlement of building foundations?
    a. Moisture level
    b. Temperature
    c. Soils types
    d. Local climate condition
    e. All of the above

39. Which of the following soils types is likely to cause differential settlement of building foundations?
    a. Clay
    b. Sandy
    c. Loam
    d. Silt-loam

40. The net movement of solvent molecules through a partially permeable membrane into a region of higher solute concentration to equalize the solute concentrations on the two sides is called?
    a. Permeability
    b. Compressibility
    c. Osmosis
    d. Cohesion

41. A project site has 60,000 square feet of land area and 12,000 square feet of gross building floor area. At a ratio of 2 square feet of parking area to 1 square foot of gross building area, how many parking spaces are required at 400 square feet per car?

42. Which of the following characteristic of an existing structure will directly affect the thermal environment of adjacent new construction? **Check the two that apply.**
    a. Albedo
    b. Texture
    c. Mechanical systems
    d. Shadow pattern

43. Which of the following is a common unit used for site area?
    a. Square yards [meters]
    b. Cubic yards [meters]
    c. Acres [hectares]
    d. Tonnage

44. An architect is working on a residential project. Which of the following are correct statements? **Check the two that apply.**
    a. The architect should review the building codes for FAR information.
    b. The architect should contact the Planning Department to obtain information on the existing easement and set back requirements for the front yard, the side yard and the back yard.
    c. The architect should contact the Planning Department to obtain site coverage ratio.
    d. The architect should contact the Planning Department to obtain zoning information for the site.

45. The most important factor for locating a supermarket is
    a. its proximity to potential clients
    b. the availability of utilities
    c. a downtown location
    d. its proximity to high income households

46. A practice or device designed to keep eroded soil on a construction site, so that it does not wash off and cause water pollution to a nearby water body is called
    a. erosion control
    b. sediment control
    c. pollution control
    d. defoliation control

47. Land use restrictions imposed by the Planning Department typically include all of the following EXCEPT **(Check the two that apply.)**
    a. setbacks
    b. CC&R
    c. height and area limitations
    d. zoning
    e. exit widths

48. All of the following are considered land use restrictions EXCEPT
    a. setbacks regulations
    b. slenderness ratio
    c. height and area limitations
    d. zoning ordinances
    e. site coverage ratio

49. In a cold climate, which of the following is the minimum requirement for a building footing?
    a. The building footing should be above frost line.
    b. The top of a building footing should align with frost line.
    c. The bottom of a building footing should align with frost line.
    d. The centerline of a building footing should align with frost line.

50. A roof overhang on which of the following façades of a building built in the southern hemisphere will provide seasonal adjustment for solar radiation?
    a. North
    b. South
    c. East
    d. West

51. Which of the following are the most important factors in the design of residential units? **Check the two that apply.**
    a. Orientation
    b. The heights and locations of adjacent buildings
    c. Bedrooms facing the dominant wind
    d. Avoiding west facing units

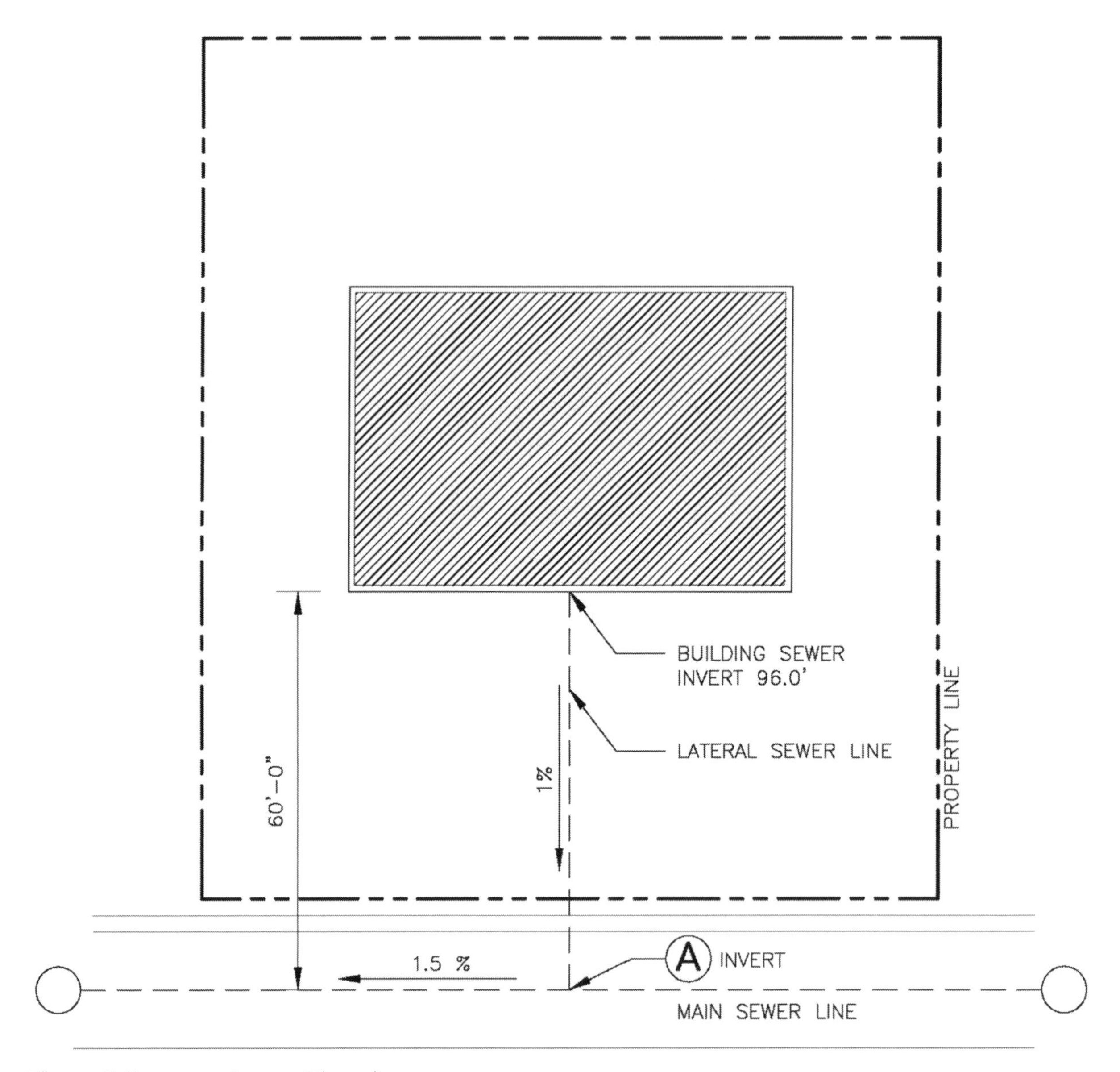

Figure 3.5 Invert Elevation

52. If the slope for the lateral sewer line is 1% and the slope for the main sewer line is 1.5%, what is the invert elevation at point A as shown on Figure 3.5?
    a. 95.1'
    b. 95.3'
    c. 95.4'
    d. 96.7'

53. Which of the following are correct? **Check the two that apply.**
    a. Detention ponds are also "dry ponds."
    b. Retention ponds are also "dry ponds."
    c. Detention ponds are used to hold stormwater for a short period of time.
    d. Retention ponds are used to hold stormwater for a short period of time.

54. High moisture content in soils can have which of the following effects? **Check the two that apply.**
    a. Uneven settlement
    b. Increased bearing capacity
    c. Heaving
    d. Decreased compatibility

55. Methods to prevent a retaining wall failure would include which of the following? **Check the four that apply.**
    a. Over-excavate below footing grade and fill with compacted gravel
    b. Drain surface water away from a retaining wall
    c. Waterproof the foundation to reduce filtration
    d. Use perforated pipe to release hydropressure
    e. Weep holes
    f. Proper footing design to overcome the turning moment
    g. Extend footings and foundations to a depth of consistent ground moisture

56. The following are considered environmental impact issues for site analysis EXCEPT **(Check the three that apply.)**
    a. reflections
    b. dominant wind direction
    c. recommended footing design
    d. archeological discoveries
    e. sun and shadow patterns
    f. demography
    g. traffic condition

57. Which of the following needs a grease interceptor? **Check the two that apply.**
    a. A single family residence
    b. A restaurant
    c. An office building
    d. A supermarket

58. Restrictive covenants are typically controlled by
    a. city
    b. contractor
    c. HOA
    d. EPA

59. Biophilia means
    a. nearsightedness
    b. farsightedness
    c. love at the first sight
    d. human beings' latent desire of being loved
    e. an instinctive bond between human beings and other living systems

60. Looking at the cost of purchasing and operating a building or product, and the relative savings is called
    a. life cycle approach
    b. life cycle assessment (LCA)
    c. life cycle analysis
    d. life cycle costing
    e. life cycle cost and saving analysis
    f. low impact development (LID)

61. Sheepsfoot is often used in
    a. excavation
    b. shoring
    c. cribbing
    d. soil compaction

62. A water district is building a new water processing plant. Which of the following is an ideal location?
    a. A location close to the lowest elevations of the city
    b. A location close to power plant
    c. A location close to residential properties
    d. A location close to commercial properties

63. An architect is reviewing a set of civil plans. If the finish floor elevation of an office building is 103', and the building has 24" wide spread footings at 5'-0" deep, then has a 6"-thick slab over 2" sand over moisture barrier over 2" sand, what should the building pad elevation on the grading plan be?
    a. 97.17'
    b. 98.0'
    c. 102.17'
    d. 102.6'

64. An architect is working on a retail building with a 25'-wide sidewalk in front of the building. If the finish floor elevation of the building is 106', the sidewalk has a 6"-high curb, and the cross slope for the sidewalk is 1.5%, what should the bottom of the curb elevation at the outside edge of the sidewalk next to the driving aisle be?
    a. 105.02'
    b. 105.12'
    c. 105.22'
    d. 105.32'

65. An architect is planning a single-family residential project near a freeway. Which of the following is the most effective way to alleviate noise from the freeway? **Check the two that apply.**
    a. Use insulated glazing for windows
    b. Plant a row of 25'-high cypress trees between the freeway and homes
    c. Plant a row of 25'-high sycamore trees between the freeway and homes
    d. Building a 12'-high CMU wall with an additional 6'-high glass screen on top.

B. Mock Exam: Site Grading Vignette

1. Directions and Code

Directions and **Code** are the same as the site grading vignette in the official NCARB exam guide. The NCARB **Directions** and **Code** has been very consistent through various versions of ARE.

You can download the official exam guide and practice program for SPD division at the following link:
http://www.ncarb.org/en/ARE/Preparing-for-the-ARE.aspx

2. Program

The program is the same as the site zoning vignette in the official NCARB exam guide, except the following (Figure 3.6):

1) The background is different.
2) We have a monument instead of a smokestack. So replace the word "smokestack" with "monument" in the program.

Note:
For your convenience, we have placed the DWG files for figure 3.6 and figure 3.7 on our website at: http://GreenExamEducation.com

Here are the simple steps for you to download the DWG file for FREE:

- *Go to http://GreenExamEducation.com/*
- *Click on "Free Downloads and Forums" on the top menu*
- *Follow the instructions on the next page.*
- *After you download the dwg files, you can install the dwg files to use with NCARB software ARE 4.0 or higher version by follow our instructions under Section D on the following pages.*

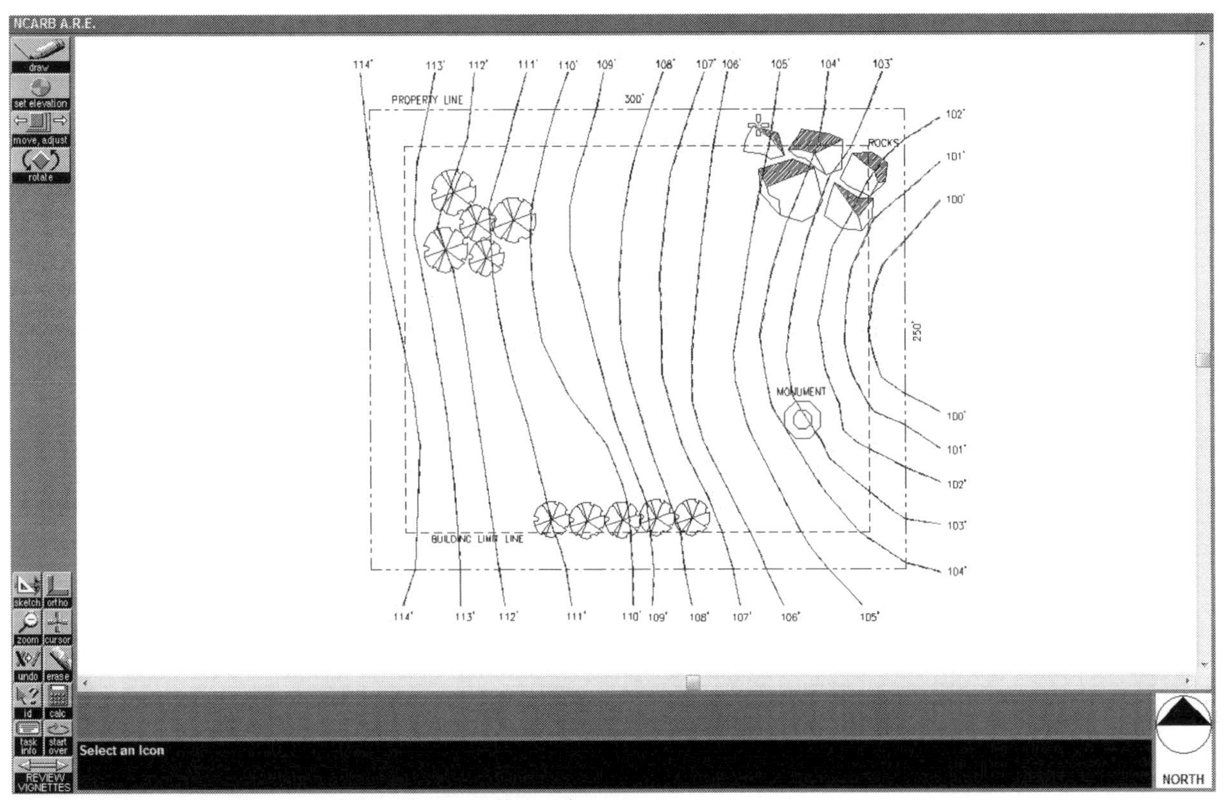

Figure 3.6 Background for site grading vignette

C. Mock Exam: Site Design Vignette

1. Directions and Code

Directions and **Code** are the same as the site design vignette in the official NCARB exam guide. The NCARB **Directions** and **Code** has been very consistent through various versions of ARE.

2. Program

The program is the same as the site design vignette in the official NCARB exam guide, except the following (Figure 3.7):

1) The background is different.
2) Draw a 9,000 ft² Pedestrian Plaza.
3) Provide only one 24 ft wide curb cut located no closer than 110 ft from the intersection of the centerlines of the two existing public streets.
4) Provide a 15 ft setback from the Pond for all construction or built improvements.

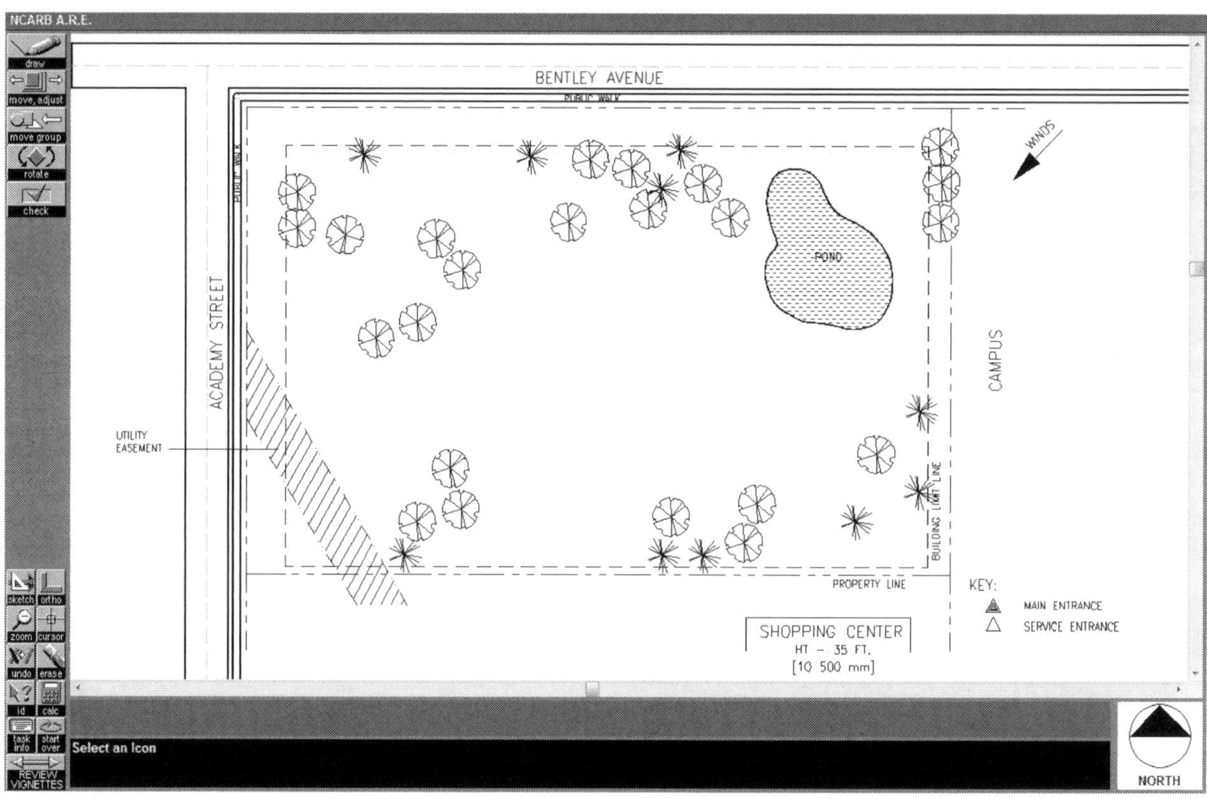

Figure 3.7 Background for site design vignette

D. Instructions on Installing Alternate DWG Files for Use with NCARB Software

1) **Right click** the **Start** button on the lower left-hand corner of your computer to open your **Windows Explore** (figure 3.8).

2) Go to the folder where you placed the downloaded DWG file. On the top pull-down menu, under **View**, select **Details**. You should see an extension for all the file names, i.e., a dot (.) followed by three letters. For example, the AutoCAD file names for our Site Planning and Design vignettes alternate drawings are **B9tut2W1_SD Mock.dwg** and **BzTUT0w1_SG Mock.dwg**; the ".dwg" is the extension (figure 3.9).

3) If you do NOT see an extension for all the file names, see the following instructions. (These directions are for Windows Vista and Window 7, but Windows XP is similar.)
   - **Windows Explore > Organize > Folder and Search Options** (figure 3.10) A menu window will pop up. In that menu, select **View** (figure 3.11).
   - You will see a list with several boxes checked. Scroll down and **uncheck** the box for **Hide extensions for known file types.**
   - Select **Apply to Folders** and a menu window will pop up. Select **Yes** to get out of the **View** menu (figure 3.12).

4) **Windows Explore > Computer > C: Drive > Program Files > NCARB.** Select the folder for the NCARB ARE division that you are working on (figure 3.13). For example, **C:\Program Files\NCARB\Programming, Planning, and Practice**

5) Open the folder and you will see files ending in .AUT and .DWG, as well as .SOL. The .AUT files are the program information, the .DWG files are the templates for the practice vignettes, and the .SOL files are your previous solution data to the official NCARB sample vignettes after you complete them. This is important.

6) Make a new folder called **Backup** under the **Site Planning and Design** folder. Copy all the .AUT files and .DWG files, as well as .SOL files to the **Backup** folder. For SPD, we have two DWG format files: **BzTUT0w1.dwg** and **B9tut2W1.dwg.**

7) Make a new subfolder called **Alts** in the **Site Planning and Design** folder. Copy the alternate DWG files **(B9tut2W1_SD Mock.dwg** and **BzTUT0w1_SG Mock.dwg)** that you want to use into the Alt folder. Rename the alternate DWG file(s) to match the name of the original NCARB DWG files. For our example, we will rename **B9tut2W1_SD Mock.dwg** as **B9tut2W1.dwg, and BzTUT0w1_SG Mock.dwg** to **BzTUT0w1.dwg** to match the original NCARB DWG file.

8) Copy the alternate DWG file(s) to the NCARB folder for your vignettes **(C:\Program Files\NCARB\ Site Planning and Design)** and overwrite the original NCARB DWG files (Figure 3.14).

9) Delete the .SOL files (**pbztut.sol** and **pb9tut.sol**) from the NCARB folder (**C:\Program Files\NCARB\ Site Planning and Design)**. Otherwise, your previous solutions will still show up when you open the NCARB software.

   Note: *NCARB practice software ARE 4.0 ONLY works with AutoCAD Release 12. If you have a DWG file that is in an AutoCAD Release 13 version or higher, the NCARB practice software ARE 4.0 will NOT work, and you have to save the DWG file as AutoCAD Release 12 file.*

   When you save the DWG file AutoCAD Release 12 file, you may lose some information such as the pen weights, the leader arrow size, etc. However, you can still read the plan and understand the concepts for this exercise.

10) Open the NCARB practice software. You may get an error message that says the DWG has been changed. Just ignore it and click OK.

11) Now you can start to work on your solution using the NCARB software.

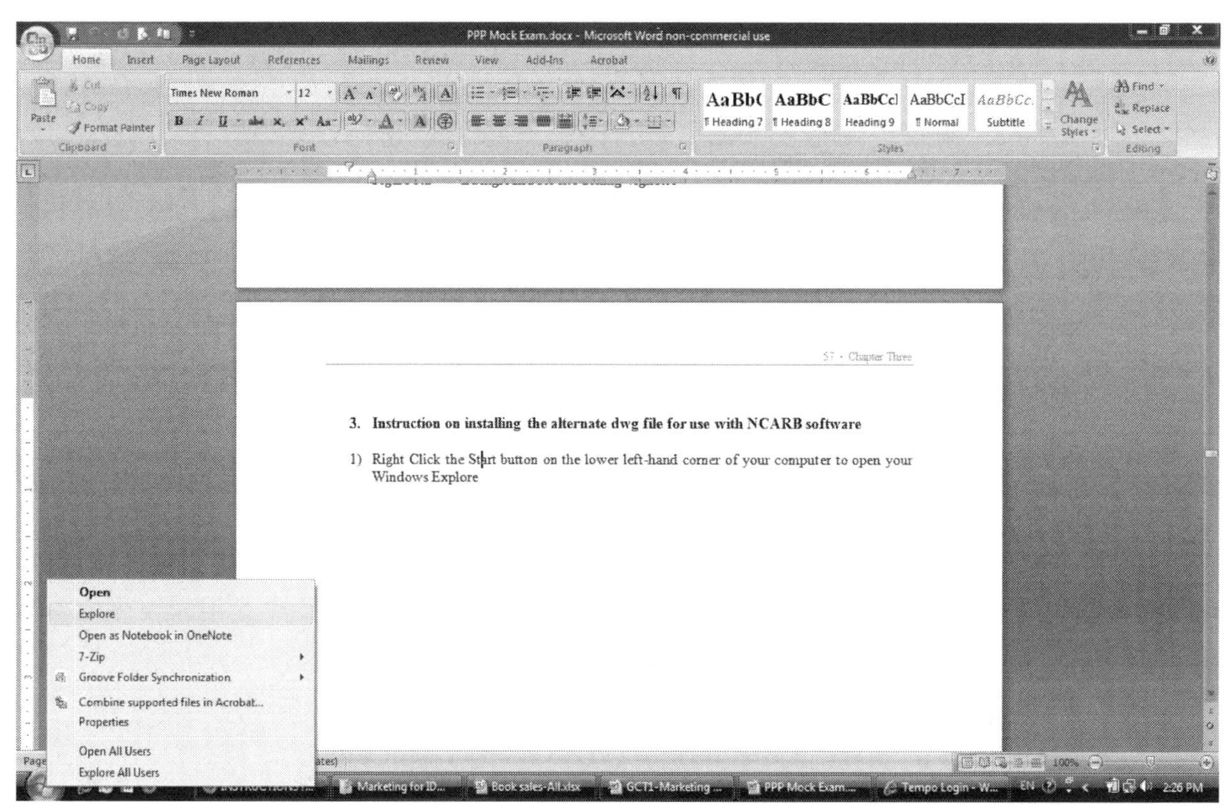

Figure 3.8 **Right click** the **Start** button on the lower left-hand corner of your computer to open **Windows Explore.**

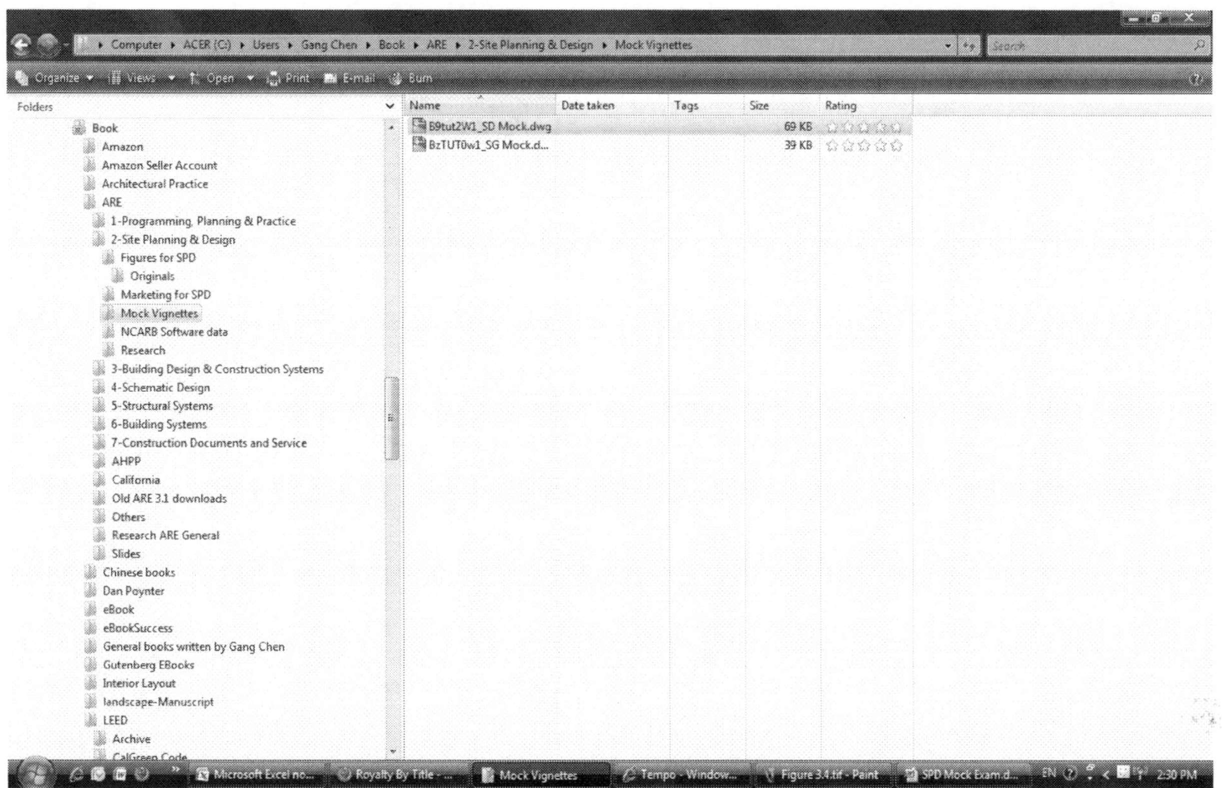

Figure 3.9 You should see an extension for all the file names, i.e., a dot (.) followed by three letters.

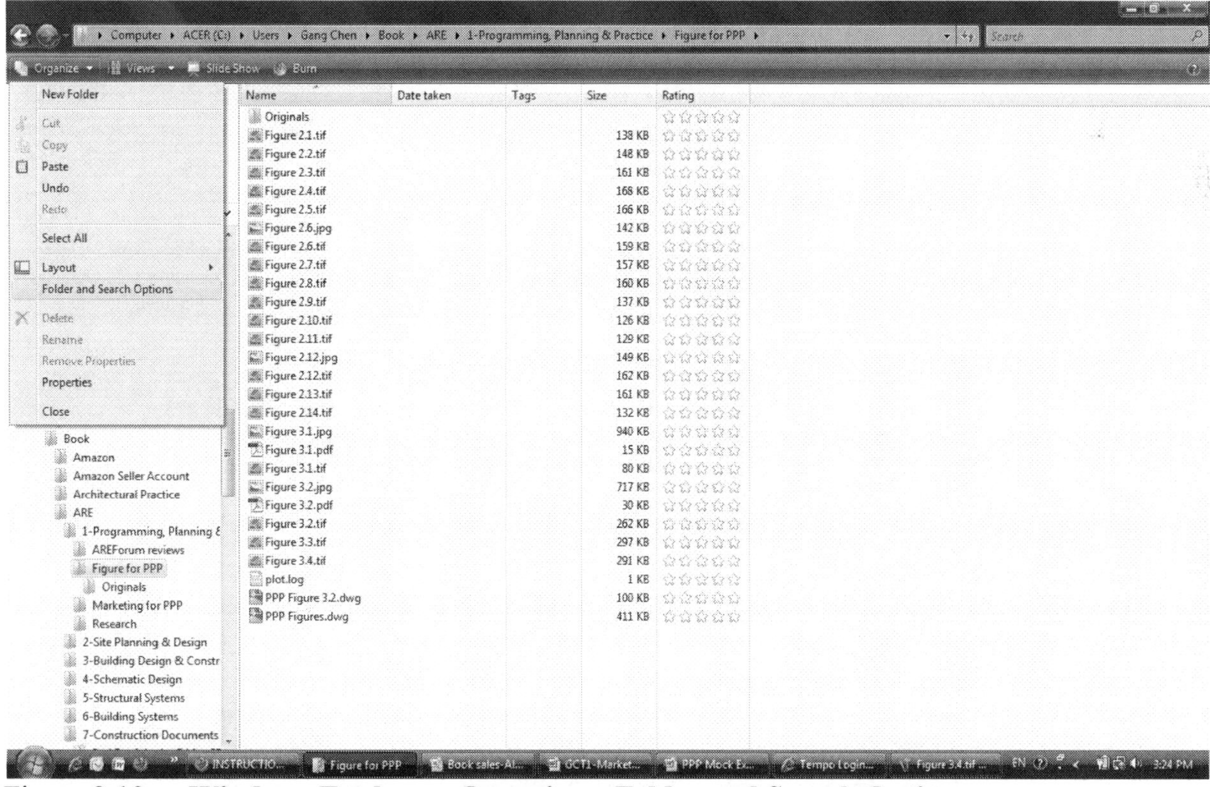

Figure 3.10 **Windows Explore > Organize > Folder and Search Options**

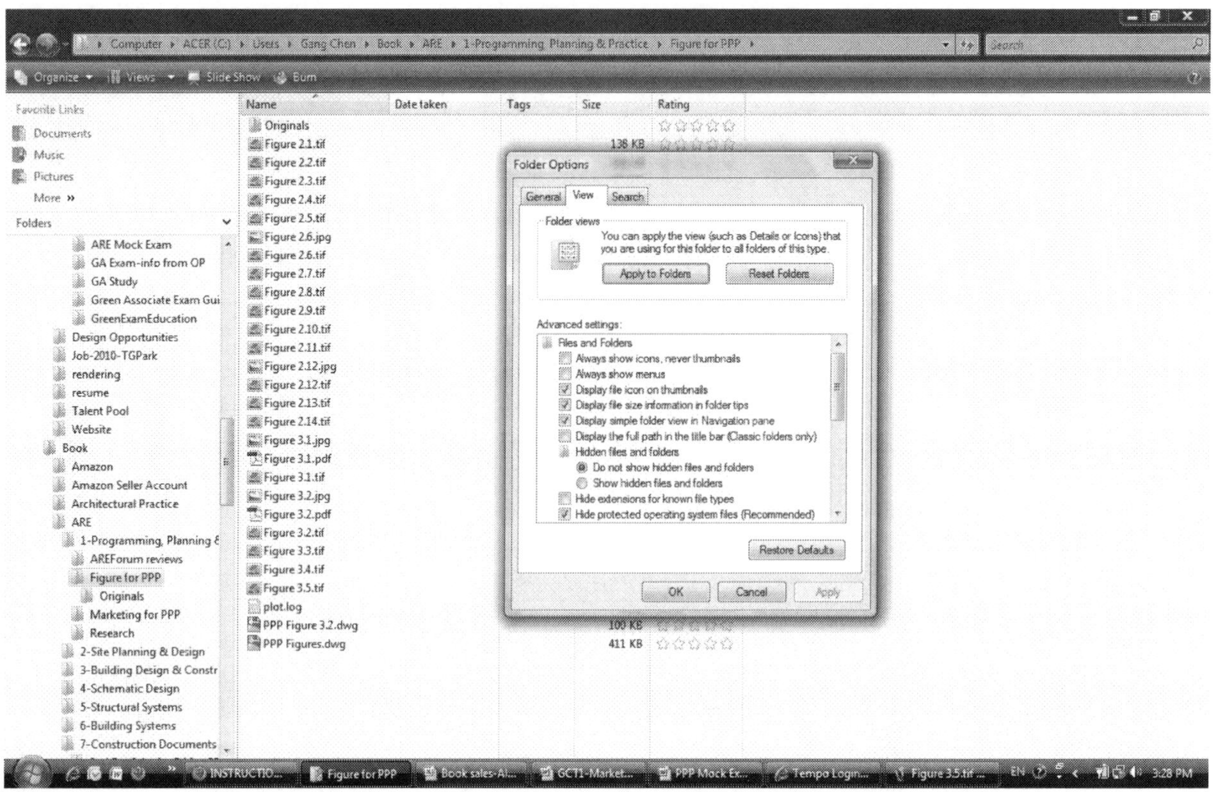

Figure 3.11 A menu window will pop up. In that menu, select **View.**

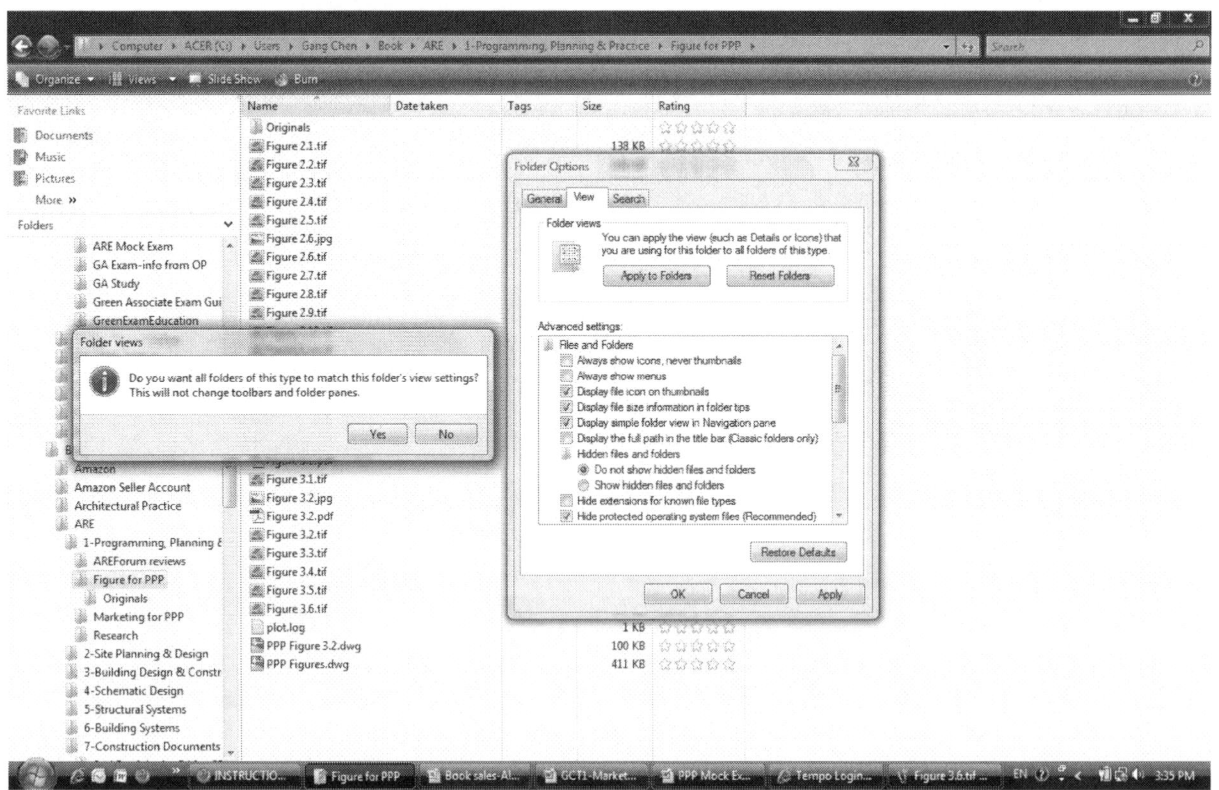

Figure 3.12 Select **Apply to Folders** and a menu window will pop up. Select **Yes** to get out of the **View** menu.

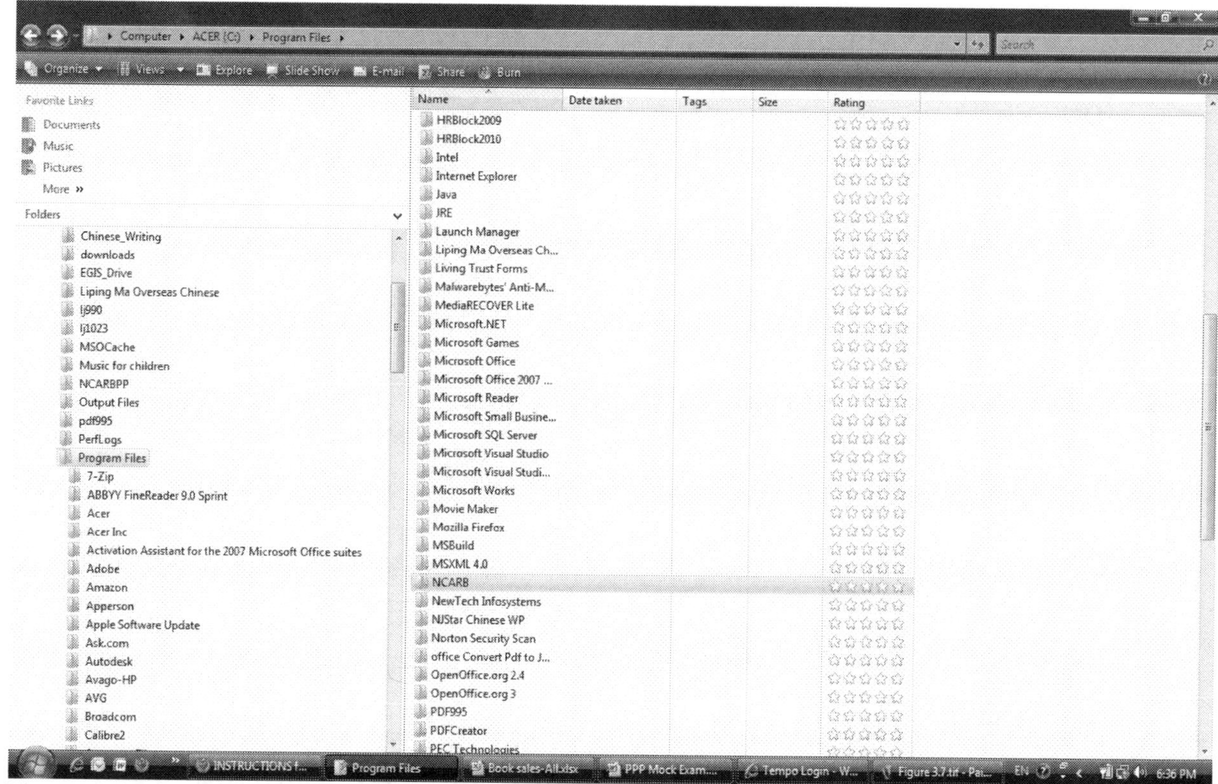

Figure 3.13 **Windows Explore > Computer > C: Drive > Program Files > NCARB.** Select the folder for the NCARB ARE division that you are working on.

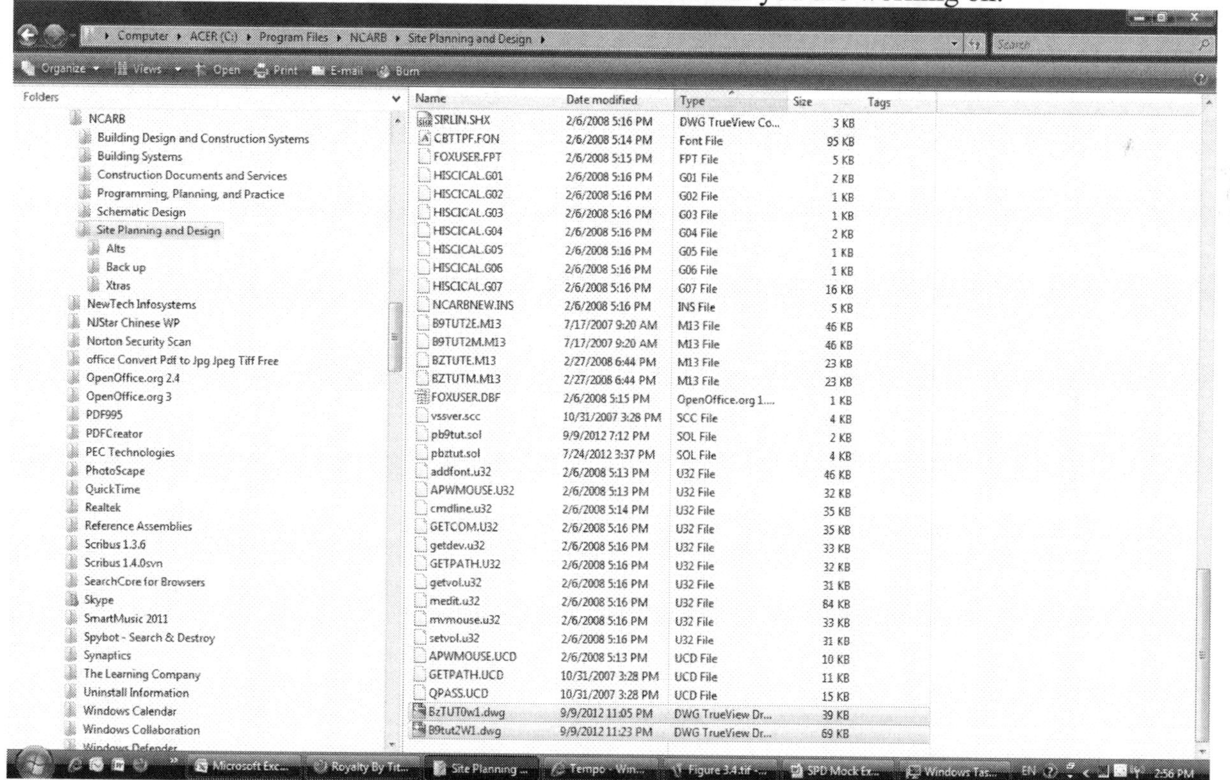

Figure 3.14 Copy the alternate DWG files to the NCARB folder for your vignettes and overwrite the original NCARB DWG files.

Chapter Four

ARE Mock Exam Solution for Site Planning & Design (SPD) Division

A. Mock Exam Answers and Explanation: SPD Multiple-Choice (MC) Section

Note: If you answer 60% of the questions correctly, you pass the MC Section of the exam.

1. Answer: f. absorption rate *of soils*
   The purpose of a **percolation test** is to determine the absorption rate *of soils* for "leach field" or a septic drain field.

   Absorption rate is a term frequently used in real estate. It is the rate at which available homes are sold in a specific real estate market during a given time period.

   Soil alkalinity is the ability of soil to neutralize acids to the equivalence point of bicarbonate or carbonate. **Alkaline soil** is soil with a pH level over 7.0.

   Soil density is the ratio of soil mass (M) to soil volume (V). An average soil density is about 100 to 110 pounds per cubic foot.

   Gravel sizes refer to the sizes of unconsolidated rock fragments.

   Evaporation rate is the rate at which a material will vaporize.

2. Answer: e. Live
   Live loads defines the force water, snow or people exert on roofs.

   Hydrostatic loads defines the force water exert on retaining walls.

   Dynamic loads are the forces that move or change when exerting on a structure, such as force of wind or a moving truck.

   Water loads and Wedge loads are **distracters** to confuse you.

3. Answer: d. D
   D is the direction of water flow because water always flows from a higher contour line to a lower contour line at a 90-degree angle, or perpendicular to the contour lines.

4. Answer: c. 112.00'
   Because the front door has a threshold, the proper spot elevation at the front entrance has to be 112.00'. Otherwise, the front entrance will exceed the maximum 1/4 inch to 1/2 inch vertical change in level and will not comply with the handicap codes.

   See cross section drawings below (Figure303.3, Bevel Change in Level):
   http://www.access-board.gov/guidelines-and-standards/buildings-and-sites/about-the-ada-standards/ada-standards/chapter-3-building-blocks

   Also see:

   http://www.ada.gov/
   AND
   http://www.access-board.gov/guidelines-and-standards/buildings-and-sites/about-the-ada-standards/ada-standards

5. Answer: b. 111.98'
   For the same one-story supermarket building in previous question, 111.96' is a proper spot elevation at the front entrance if the front door has no threshold. This is because 1/4 inch (or 0.02') is the maximum vertical change in level: 112.00'-0.02' = 111.98'.

   We set the spot elevation to as low as possible per the handicap codes to prevent water from getting into the building. The site walk near the front entrance also needs to slope away from the building at 1% to 2%. We suggest setting the slope at 1.5%.

6. Answer: d. Gravel
   Figure 3.2 shows a typical retaining wall detail with gravels around a perforated pipe. This is to release the hydrostatic pressure behind the retaining all. Neither sand nor soils are proper materials around a perforated pipe because they are too fine and will leak into the perforated pipe.

7. Answer: c. Perforated pipe
   B as shown on Figure 3.2 is perforated pipe. See explanation in answer to previous question.
8. Answer: c. 113'
   The spot elevation = 115'- 5' x (10'/25') = 115' - 2' = 113'

9. Answer: a and d
   For a ***convex* slope**, the contour lines are closer to each other at the *lower* portion of the slope. A ***convex* slope** has a section profile as shown in Section D, the section of Plan A. Section C is the section of Plan B.

10. Answer: b and c

For a ***concave* slope,** the contour lines are closer to each other at the *higher* portion of the slope.

It is very easy to confuse the term convex slope with the term concave slope. Here is a simple **mnemonic** to help you differentiate them:

A *concave* slope contains a cave and has a "dent."

11. Answer: b and c

The following is true:

- In Southern Hemisphere, the solar altitude angle is highest in December and lowest in June.
- In Northern Hemisphere, the solar altitude angle is lowest in December and highest in June.

12. Answer: d

The following is the most energy efficient orientation of a rectangular building in Los Angeles:

- Locate the building with the long direction oriented along the east-west axis and facing slightly to the west.

This is because Los Angeles is located in the subtropical area of the Northern Hemisphere, and the dominant wind in summer is from southwest. Locating the building with the long direction oriented along the east-west axis will minimize excess heat gain and maximize the use of the cool ocean breeze and dominant wind from southwest in summer.

13. Answer: a and d

As an architect, you will probably do at least one residential remodel or room addition project in your lifetime. This problem addresses an important issue for you to consider for a residential remodel or room addition project.

The following are correct statements:

- The architect *can* connect the new toilets to the existing 4" sewer line.

In fact, the architect *has to* connect the new toilets to the existing 4" sewer line: She can add another 3" lateral sewer line to connect the new toilets to the existing 4" sewer line or upgrade the existing 3" interior sewer line to 4".

Newer single-family homes typically have 4" sewer line. However, some older homes still have 3" sewer lines. There are detailed tables on sizing the sewer lines, and a 3" sewer line is definitely NOT adequate for the loads of 4 toilets, and the architect *has to* connect the new toilets to the existing 4" sewer line.

- The architect *can* ask his plumbing engineer for advice.

This is not an ideal answer but it is still correct. Of course, the architect *can* ask his plumbing engineer for advice if she has one. For a residential remodel or room addition project, electrical, mechanical and plumbing is typically designed and built by the contractor.

The following are incorrect statements:

- The architect can connect the new toilets to the existing 3" sewer line.
- It is better to oversize a sewer line when designing buildings.
  It is not a good idea to oversize a sewer line when designing buildings because an oversized sewer line does not have enough water to carry the waste, and gets clogged more often than a properly-sized sewer line.

14. Answer: b. 1:12

A typical slope of the flared side shall not exceed 1:10. However, if the landing depth at the top of a curb ramp is less than 48 inches, then the slope of the flared side shall not exceed 1:12.

See cross figure drawing showing curb ramps flared sides at (Figure 406.3 Sides of Curb Ramps):
http://www.access-board.gov/guidelines-and-standards/buildings-and-sites/about-the-ada-standards/ada-standards/single-file-version#a4

Also see:

http://www.ada.gov/

AND
http://www.access-board.gov/guidelines-and-standards/buildings-and-sites/about-the-ada-standards/ada-standards

15. Answer: b. 36"
Per *International Building Code* (IBC),Section 1010.6, the minimum width of a means of egress ramp between handrails, if provided, or other permissible projections shall be 36".

See *International Building Code* (IBC),Section 1010.6, at the following link:
http://publicecodes.cyberregs.com/icod/ibc/2012/icod_ibc_2012_10_sec010.htm

SPD is a very broad ARE division, and you may get a few ARE exam questions from other ARE divisions. That is why we throw in a few questions from other ARE divisions in the SPD mock exam also.

16. Answer: c. trees
Trees are unlikely to show up on an ALTA survey plan.

ALTA means **American Land Title Association.** It is the national trade association of the **abstract and title insurance industry** and was founded in 1907.

ALTA's official website is:
http://www.alta.org/

My other book, ***Building Construction:*** *Project Management, Construction Administration, Drawings, Specs, Detailing Tips, Schedules, Checklists, and Secrets Others Don't Tell You (Architectural Practice Simplified, 2nd edition)*, has detailed discussions on related terms such as ALTA, CALTA and ACSM.

17. Answer: c
    The following is the correct order for arranging the units used in the Public Land Survey System (PLSS) in the US, from large to small.
    - check, township, section

    A **check** is a square parcel of land 576 square miles in area with 24-mile-long sides.
    A **township** is a square parcel of land 36 square miles in area with 6-mile-long sides.
    A **section** is a square parcel of land 1 square mile in area with 1-mile-long sides.

18. Answer: a
    She should submit a Conditional Use Permit application to the city.
    - **Non-conforming Use**: the use that does not conform to the current zoning ordinance, but allowed to stay or grandfathered in.
    - **Incentive Zoning**: Use the permission to build a larger or taller building as an incentive for a private developer to provide public amenities such as a park.
    - **Ordinance Variance**: This is just a **distracter**.

19. Answer: a and d

    Albedo is a word derived from Latin *albedo* "whiteness" (or reflected sunlight). It means the ratio of reflected sunlight. The albedo of a perfectly white surface is 1, and the albedo of a perfectly black surface is 0.

    The following statements are correct:
    - In a cold climate, an architect should use materials with low albedo and low conductivity for ground surfaces. These surfaces will absorb more sunlight and retain the heat as close to the surfaces as possible.
    - In a tropical climate, an architect should use materials with high albedo and high conductivity for ground surfaces. These surfaces will absorb less sunlight and quickly absorb and store the heat, and as quickly released when the temperature drops. This will help to produce a mild and stable microclimate.

    The following statements are incorrect:
    - In a cold climate, an architect should use materials low with albedo and high conductivity for ground surfaces.

- In a tropical climate, an architect should use materials with low albedo and low conductivity for ground surfaces.

20. Answer: Per EPA, using a watershed-based approach to wetland protection ensures that the whole system, including land, air, and water resources, is protected.

    See *Wetland Overview* from the EPA:
    http://water.epa.gov/type/wetlands/upload/2005_01_12_wetlands_overview.pdf

21. Answer: a, b, d, and f
    Per EPA, the following are general categories of wetlands found in the US:
    - Bogs
    - Fens
    - Marshes
    - Swamps

    Lakes and Reservoirs are just distracters.

    According to EPA:
    "**Marshes** are wetlands dominated by soft-stemmed vegetation, while **swamps** have mostly woody plants. **Bogs** are freshwater wetlands, often formed in old glacial lakes, characterized by spongy peat deposits, evergreen trees and shrubs, and a floor covered by a thick carpet of sphagnum moss. **Fens** are freshwater peat-forming wetlands covered mostly by grasses, sedges, reeds, and wildflowers."

    See *Wetland Overview* from the EPA for more information:
    http://water.epa.gov/type/wetlands/upload/2005_01_12_wetlands_overview.pdf

22. Answer: c
    Per EPA, habitat degradation is the leading cause of species extinction.

    Human activities, pollution and excessive hunting all play a role in species extinction, but they are not the leading cause.

23. Answer: c
    SWPPP stands for Stormwater Pollution Prevention Plan.

    The following are all **distracters**:
    - Solid Waste Pollution Prevention Plan
    - Sewer Waste Pollution Prevention Plan
    - Soiled Water Pollution Prevention Plan

    See *Developing your Stormwater Pollution Prevention Plan: A Guide for Construction Sites* for more info:
    http://www.epa.gov/npdes/pubs/sw_swppp_guide.pdf

24. Answer: f. none of the above

    Pay attention to the word "not" in the original question.

    The following are the purposes of SWPPP, and therefore the incorrect answers:
    - to prevent stormwater contamination
    - to prevent airborne dust from getting into rivers
    - to control sedimentation and erosion
    - to comply with the requirements of the Clean Water Act
    - to comply with the requirements of CGP (construction general permit)

    See *Developing your Stormwater Pollution Prevention Plan: A Guide for Construction Sites* for more info:
    http://www.epa.gov/npdes/pubs/sw_swppp_guide.pdf

25. Answer: a. National Pollutant Discharge Elimination System

    The following are all **distracters**:
    - National Pollutant Discharge and Exchange System
    - Natural Pollutant Discharge Elimination System
    - Natural Pollutant Discharge and Exchange System

    See *Developing your Stormwater Pollution Prevention Plan: A Guide for Construction Sites* for more info:
    http://www.epa.gov/npdes/pubs/sw_swppp_guide.pdf

26. Answer: c. Civil Engineer

    The owner or the architect may request a Best Management Practices (BMP) plan, and the contractor will execute a Best Management Practices (BMP). However, a civil engineer is typically in charge of preparing a Best Management Practices (BMP) plans.

27. Answer: a, c, d, and e

    The following are structural BMPs:
    - Fences
    - Sedimentation ponds
    - Erosion control blankets
    - Temporary or permanent seeding,

    The following are non-structural BMPs:
    - Maintaining equipment

- Training site staff on erosion and sediment control practices.

Per *Developing your Stormwater Pollution Prevention Plan: A guide for Construction Sites*:
"**Structural BMPs** include silt fences, sedimentation ponds, erosion control blankets, and temporary or permanent seeding, while **non-structural BMPs** include picking up trash and debris, sweeping up nearby sidewalks and streets, maintaining equipment, and training site staff on erosion and sediment control practices."

28. Answer: b and d
Please note we are looking for *incorrect* statements.

The following are incorrect statements, and therefore the correct answers:
- It is usually easier and less expensive to control sedimentation than it is to prevent erosion.
- A good SWPPP will use one kind of BMPs for the best results.

The following are correct statements, and therefore the incorrect answers:
- A good SWPPP will use both kinds of BMPs in combination for the best results.
- It is usually easier and less expensive to prevent erosion than it is to control sedimentation.

29. Answer: d
Pay attention to the word "not" in the original question.

Plumbing plans are typically not included in a SWPPP.

The following are typically included in a SWPPP:
- Site plans
- Maintenance/inspection procedures
- Description of controls to reduce pollutants

See *Developing your Stormwater Pollution Prevention Plan: A Guide for Construction Sites* for more info:
http://www.epa.gov/npdes/pubs/sw_swppp_guide.pdf

30. Answer: a and c
The following are objectives of SWPPP:
- *Reduce* impervious surfaces and promote infiltration
- Protect ***surface* waters** from erosion and sediment

The following are not objectives of SWPPP:
- *Increase* impervious surfaces and promote infiltration
- Protect *groundwater* from erosion and sediment (**Groundwater** is water located beneath the earth's surface. SWPPP can help to recharge *groundwater* but no direct impact on protecting *groundwater* from erosion and sediment).

31. Answer: d
    Fingerprinting your site means inventorying a site's natural features.

    The following are all **distracters**:
    - Obtaining the fingerprint of everyone at your site
    - Obtaining the fingerprint of every visitor to your site
    - Identifying the trees in your site

32. Answer: a and c
    The following is true regarding SWPPP:
    - Silt fencing is more effective than inlet protection
    - Hydromulching is more effective than inlet protection

    The following is untrue regarding SWPPP:
    - Silt fencing is less effective than inlet protection
    - Hydromulching is less effective than inlet protection

    See *Developing your Stormwater Pollution Prevention Plan: A Guide for Construction Sites* for more info:
    http://www.epa.gov/npdes/pubs/sw_swppp_guide.pdf

33. Answer: e
    The construction site operator must sign the SWPPPs and inspection reports. The construction site operator can be the general contractor, or a duly authorized representative of the general contractor. S/he may not be the concrete contractor.

    According to *Developing your Stormwater Pollution Prevention Plan: A Guide for Construction Sites*:
    "All reports, including SWPPPs and inspection reports, generally must be signed by the construction site operator or a duly authorized representative of that person."

    The following are incorrect answers:
    - Owner
    - Architect
    - Civil Engineer

34. Answer: a and c
    The following is true regarding Runoff Coefficient:
    - Downtown areas tend to have a higher Runoff Coefficient than residential areas.
    - Drives and walks tend to have a higher Runoff Coefficient than lawns.

    The following are incorrect answers:
    - Downtown areas tend to have a lower Runoff Coefficient than residential areas.
    - Drives and walks tend to have a lower Runoff Coefficient than lawns.

See Table C-1 on *Developing your Stormwater Pollution Prevention Plan: A Guide for Construction Sites*:
http://www.epa.gov/npdes/pubs/sw_swppp_guide.pdf

35. Answer: b
Pay attention to the word "not" in the original question.

Biocontrol is not a groundwater remediation technology, and therefore the correct answer.

The following are groundwater remediation technologies, and therefore the incorrect answers:
- Bioventing
- Pump and treat
- Air sparging

Bioventing uses in situ (on-site) microorganisms to biodegrade organic constituents in the groundwater.

Pump and treat is a process of pumping water to the surface and mixing it with either biological or chemical treatments to remove the impurities.

Air sparging blows air directly into the ground water and creates bubbles to remove contaminants from the groundwater. The bubbles carry the contaminants up into the unsaturated zone (i.e., soil). As the contaminants move into the soil, a soil vapor extraction system removes vapors.

36. Answer: c
The following is the correct order of senses in conveying information about a site, arranged from most important to least important:
- Sight, hearing, smell, touch, taste

The following are incorrect answers:
- Sight, touch, hearing, smell, taste
- Sight, hearing, touch, smell, taste
- Sight, touch, smell, hearing, taste

37. Answer: b and c
Pay attention to the word "not" in the original question.

The following is generally NOT given in a soils report, and therefore the correct answers:
- Slump test results (**Slump test** is a test to measure the workability of fresh concrete.)
- Structural calculations

The following are incorrect answers:
- Recommended types of foundations
- Soils bearing capacity

- Type of concrete to use

38. Answer: e. All of the above
    This question tests your patience and trains you to be patient enough to finish reading a question and ALL of the choices.

    All the following are likely to cause differential settlement of building foundations.
    - Moisture level
    - Temperature
    - Soils types
    - Local climate condition

39. Answer: a
    Clay is most likely to cause differential settlement of building foundations because clay is the most expansive soil.

    Loam is the soil composed of roughly 40% sand, 40% silt, and 20% clay.
    .
    The following soils types are incorrect answers:
    - Sandy
    - Loam
    - Silt-loam

40. Answer: c
    The net movement of solvent molecules through a partially permeable membrane into a region of higher solute concentration to equalize the solute concentrations on the two sides is called **osmosis**.

    Permeability measures the ease with which a particular fluid flows through the voids of a soil.

    Compressibility measures the relative volume change of a solid or fluid in response to a pressure change.

    Cohesion is the intermolecular attraction between like-molecules.

41. Answer:

60 parking space

12,000 square feet of gross building floor area x 2 = 24,000 square feet of parking area

24,000 square feet of parking area/ 400 square feet per car = 60 parking space

The 60,000 square feet of land area in the original question is a **distracter** to confuse you.

42. Answer: a and d

The following characteristic of an existing structure will directly affect the thermal environment of adjacent new construction:

- **Albedo**: it will affect the amount of sunlight reflected to the adjacent new construction, and affect its thermal environment.
- Shadow pattern

The following are incorrect answers:

- Texture
- Mechanical systems

43. Answer: c

Acres [hectares] is a common units used for site area.

The following are incorrect answers:

- Square yards [meters]
- Cubic yards [meters]
- Tonnage

44. Answer: c and d

An architect is working on a residential project. The following are correct statements:

- The architect should contact the Planning Department to obtain site coverage ratio.
- The architect should contact the Planning Department to obtain zoning information for the site.

The following are incorrect answers:

- The architect should review the building codes for FAR information (She should contact the Planning Department to obtain FAR information).
- The architect should contact the Planning Department to obtain information on the existing easement and set back requirements for the front yard, the side yard and the back yard (She should contact the owner for easement information. The owner should have received a title report from the title company, and the title report should have the easement information).

45. Answer: a
    The most important factor for locating a supermarket is its proximity to potential clients.

    Both a and d are potential correct answers, but answer d (its proximity to high income households) is not as good as answer a (its proximity to potential clients) because the original question does not mention this is a high-end or low-end supermarket, and people from high income households may NOT shop at this supermarket.

    The following are incorrect answers:
    - the availability of utilities (this is a secondary consideration when compared with others)
    - a downtown location (it may not be a good choice because it may not have large enough space for a supermarket, and it may not provide proximity to potential clients)

46. Answer: b
    A practice or device designed to keep eroded soil on a construction site, so that it does not wash off and cause water pollution to a nearby water body is called **sediment control**. Sediment control is preventing the *disturbed* soils from entering a nearby water body. Sediment basins and silt fences are examples of sediment control.

    Erosion control is the practice of controlling wind or water erosion in land development, agriculture, and construction. Erosion control is *holding* the soils *in place*. It often involves a physical barrier, such as vegetation or rock.

    The following are incorrect answers:
    - Pollution control
    - Defoliation control
    - erosion control

47. Answer: b and e
    Pay attention to the word "EXCEPT."

    Land use restrictions imposed by the Planning Department typically include all of the following EXCEPT CC&R and Exit widths:
    - Set backs
    - Height and area limitations
    - Zoning

    Covenant Conditions and Restrictions (CC&R) is imposed by Home Owners' Association (HOA).

    Exit widths are imposed by building codes.

48. Answer: b
All of the following are considered land use restrictions EXCEPT slenderness ratio.
- Setbacks regulations
- Height and area limitations
- Zoning ordinances
- Site coverage ratio

Slenderness ratio is a structural requirement for columns or high-rise buildings.

49. Answer: c
Pay attention to the word "minimum."

In a cold climate, the following is the *minimum* requirement for a building footing:
- The *bottom* of a building footing should align with frost line.

The following are incorrect answers:
- The building footing should be above frost line. (This is incorrect because the footing is not deep enough to prevent heaving)
- The top of a building footing should align with frost line. (This is a conservative approach and works, but it is not the *minimum* requirement.)
- The centerline of a building footing should align with frost line.(This is also a conservative approach and works, but it is not the *minimum* requirement.)

50. Answer: a
Please note this question is regarding the *southern* hemisphere.

A roof overhang on which of the *north* façades of a building built in the *southern* hemisphere will provide seasonal adjustment for solar radiation.

If the question is regarding the *northern* hemisphere, then the answer should be b, South (façade).

51. Answer: a and b
The following are the most important factors in the design of residential units because they are important for ensuring each unit can receive sun for at least part of a winter day:
- Orientation
- The heights and locations of adjacent buildings

The following are incorrect answers:
- Bedrooms facing the dominant wind (It is important, but not the most important factors among the four choices).
- Avoiding west facing units (It is important, but not the most important factors among the four choices).

52. Answer: c
    If the slope for the lateral sewer line is 1% and the slope for the main sewer line is 1.5%, the invert elevation at point A as shown on Figure 3.5 is 95.4'.
    60' x 1% =0.6
    96.0 - 0.6 = 95.4'

    The following is unnecessary information used as a **distracter** to confuse you:
    The slope for the main sewer line is 1.5%.

53. Answer: a and c
    The following are correct:
    - **Detention ponds** are also "dry ponds."
    - **Detention ponds** are used to hold stormwater for a short period of time.

    Retention ponds are "wet ponds." They are artificial lakes with vegetation around the perimeter.

54. Answer: a and c
    High moisture content in soils can have the following effects:
    - Uneven settlement
    - Heaving

    The following are incorrect answers:
    - Increased bearing capacity
    - Decreased compatibility

55. Answer: b, d, e, and f
    Methods to prevent a retaining wall failure would include the following:
    - Drain surface water away from a retaining wall
    - Use perforated pipe to release hydropressure
    - Weep holes
    - Proper footing design to overcome the turning moment

    The primary methods to prevent a retaining wall failure are to release the hydropressure and to use proper footing design to overcome the turning moment.

    The following are incorrect answers:
    - Waterproof the foundation to reduce filtration (This helps to keep moisture from entering building, but is not effective in preventing a retaining wall failure).
    - Over-excavate below footing grade and fill with compacted gravel (This helps to prevent the uneven settlement of a building, but is not effective in preventing a retaining wall failure).
    - Extend footings and foundations to a depth of consistent ground moisture (This helps to prevent the uneven settlement of a building, but is not effective in preventing a retaining wall failure).

56. Answer: c, d and f
Pay attention to the words "EXCEPT."

The following are NOT considered an environmental impact issue for site analysis, and therefore the correct answers:
- Recommended footing design
- Archeological discoveries
- Demography

The following are considered an environmental impact issue for site analysis, and therefore the incorrect answers:
- Reflections
- Dominant wind direction
- Sun and shadow patterns
- Traffic condition

57. Answer: b and d
The following need a grease interceptor because they involve foods and generate grease:
- A restaurant
- A supermarket

A grease interceptor can intercept grease and prevent grease from entering main sewer line.

The following are incorrect answers:
- A single family residence
- An office building

58. Answer: c
Restrictive covenants are typically controlled by HOA, or Home Owners' Association.

The following are incorrect answers:
- City
- Contractor
- EPA

59. Answer: e
Biophilia means an instinctive bond between human beings and other living systems.

The following are incorrect answers:
- Nearsightedness
- Farsightedness
- Love at the first sight
- Human beings' latent desire of being loved

60. Answer: d
    Looking at the cost of purchasing and operating a building or product, and the relative savings is called **life cycle costing**.

    Life cycle approach: Looking at a product or building through its entire life cycle.

    Life cycle assessment (LCA): Use life cycle thinking in environmental issues.

    Low impact development (LID): A land development approach mimicking natural systems and managing storm water as close to the source as possible.

    Life cycle analysis: a technique to assess environmental impacts associated with all the stages of a product's or life cycle.

    Life cycle cost and saving analysis is an invented term used as a distracter to confuse you.

61. Answer: d
    Sheepsfoot is a tamper roller often used in soil compaction.

    The following are incorrect answers:
    - Excavation
    - Shoring
    - Cribbing

62. Answer: a
    A water district is building a new water processing plant. The following is an ideal location:
    - A location close to the lowest elevations of the city (This location makes it possible to use gravity, the most dependable natural force, for the sewer system, and substantially reduce or even eliminate the use of power).

    The following are incorrect answers:
    - A location close to power plant
    - A location close to residential properties
    - A location close to commercial properties

63. Answer: c
    The building pad elevation on the grading plan should be 102.17'.

    Building pad elevation = finish floor elevation – slab thickness – the thickness of both layers of sand = 103' - 6"- 2"- 2" = 102.17'

64. Answer: b
The bottom of the curb elevation at the outside edge of the sidewalk next to the driving aisle should be 105.12'

The bottom of the curb elevation at the outside edge of the sidewalk next to the driving aisle = the finish floor elevation of the building – elevation change because of the cross slope – the height of the curb = 106'- (25' x 1.5%) - 6" = 106'- 0.38' – 0.5' = 105.12'.

65. Answer: a and d
The following is the most effective ways to alleviate noise from the freeway:
- Use insulated glazing for windows
- Building a 12'-high CMU wall with an additional 6'-high glass screen on top.

The following are incorrect answers because even dense plants are not an effective sound barrier:
- Plant a row of 25'-high cypress trees between the freeway and homes
- Plant a row of 25'-high sycamore trees between the freeway and homes

B. Mock Exam Solution: Site Grading Vignette

1. A Step-by-step solution to the graphic vignette: Site Grading Vignette

1) **Overall concept:**
   - The Locomotive Display needs to be level: This means you need to have a contour line of the same elevation wrap around the Display.
   - Regrade the site so that water will flow around and away from the Locomotive Display: This means you need to create 2 swales and wrap them around the Display.
   - The slope of the regraded portions of the site shall be at least 2% and no more than 20%: Since the vertical elevation difference between the adjacent contour lines is 1', this means the horizontal distance between the adjacent contour lines has to be between 5' and 50'.

   Note:
   Slope = (difference of adjacent contour elevations/ horizontal distance between the adjacent contour lines)

   Horizontal distance between the adjacent contour lines = (difference of adjacent contour elevations/slope)

   1'/20% = 5'
   1'/2% = 50'

2) Use **Draw > Locomotive Display** to draw a Locomotive Display within the building limit line (Figure 4.1). Leave about the same space between the Display and the building limit line, and between the Display and the existing rocks.

3) Use **Sketch > Circle** to pre-draw a dozen 5'-diameter (OR 2'6"-radius) circles for checking the maximum slope later. Once you draw one 5'-diameter circle, and then each time you click, the software will automatically draw another 5'-diameter circle (Figure 4.2)

4) Use **Sketch > Line** to draw the centerlines of the two swales. The two swales should start on the high side of the Locomotive Display, and wrap around the sides of the Display to divert the water (Figure 4.3).

5) Click on **Move, Adjust**, and the contour lines will become highlighted (Figure 4.4).

6) Click on the contour lines (NOT necessarily right on the small squares) to adjust them to form the two swales (Figure 4.5).

   Note:
   - *We try a different strategy this time: The existing 107' contour line is the existing contour line that hits the middle of the Display; we select it as the contour line that wraps around the Locomotive Display. This way, the elevation of the Display will be 107'-6", the **cut and fill** amount of the soils will be roughly equal, and we can reduce labor of earthwork and related construction cost.*

- *Start by adjusting the 107' contour line that wraps around the Locomotive Display, and then adjust other contour lines accordingly.*
- *The direction of the water flow is opposite to the "arrow" formed by a swale's contour lines.*
- *Use the pre-drawn 5'-diameter circles to check and make sure the horizontal distance between the adjacent contour lines is larger than 5' (less than 20% slope). Adjust the contour lines if necessary (Figure 4.6).*
- *Use sketch line and id tool to check and make sure the horizontal distance between the adjacent contour lines is less than 50' (larger than 2% slope). Adjust the contour lines if necessary (Figure 4.7).*
- *The slope between the Locomotive Display and the contour line that wraps around the Locomotive Display is not subject to the maximum slope requirement, so place the contour line as close to the Locomotive Display as possible. Adjust the contour lines if necessary (Figure 4.8).*
- *Make sure the contours around the smoke stack, the rock and the trees are NOT disturbed.*

7) Use Zoom to zoom out, and click on the **Set Elevation** button on the left-hand side menu. A dialog box appears. Use the **up and down arrow** to set the elevation to 107'-6". This is 6" higher than the 107' contour line that wraps around the Locomotive Display (Figure 4.9).

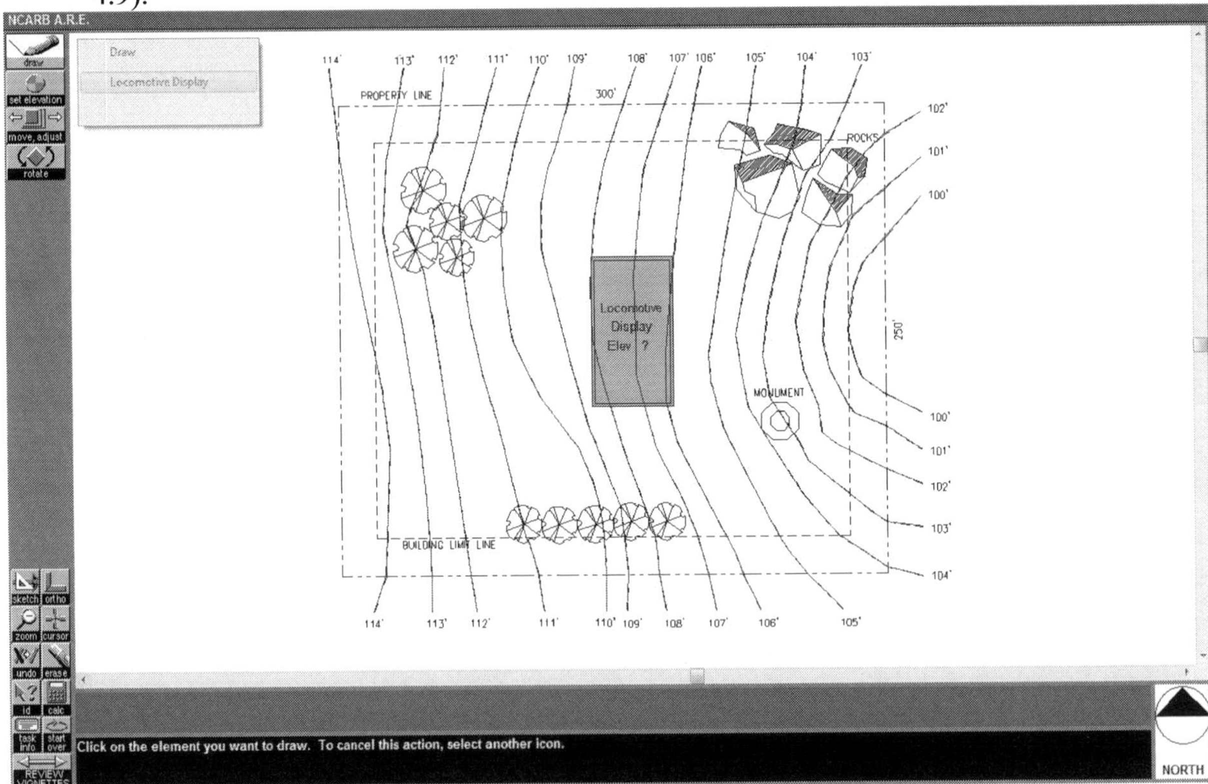

Figure 4.1 Use **Draw > Locomotive Display** to draw a Locomotive Display within the building limit line.

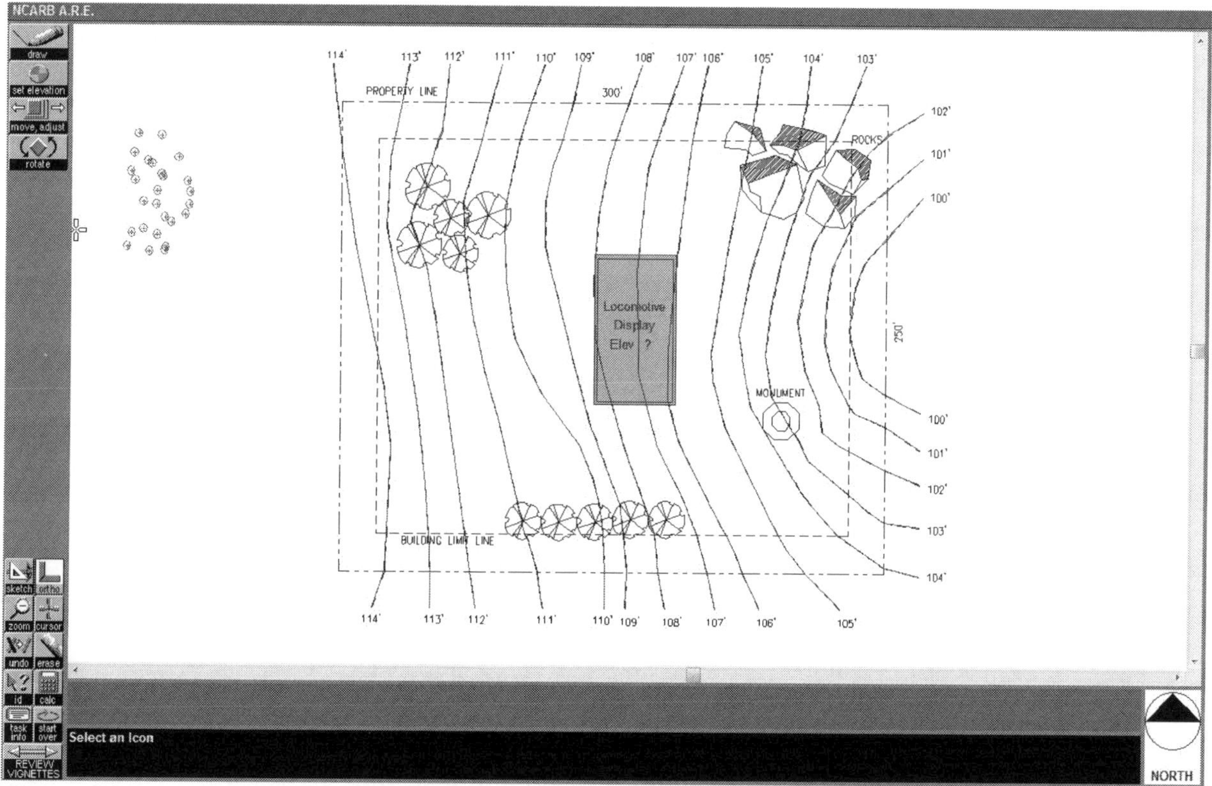

Figure 4.2 Use **Sketch > Circle** to pre-draw a dozen 5'-diameter (OR 2'6"-radius) circles for checking the maximum slope later.

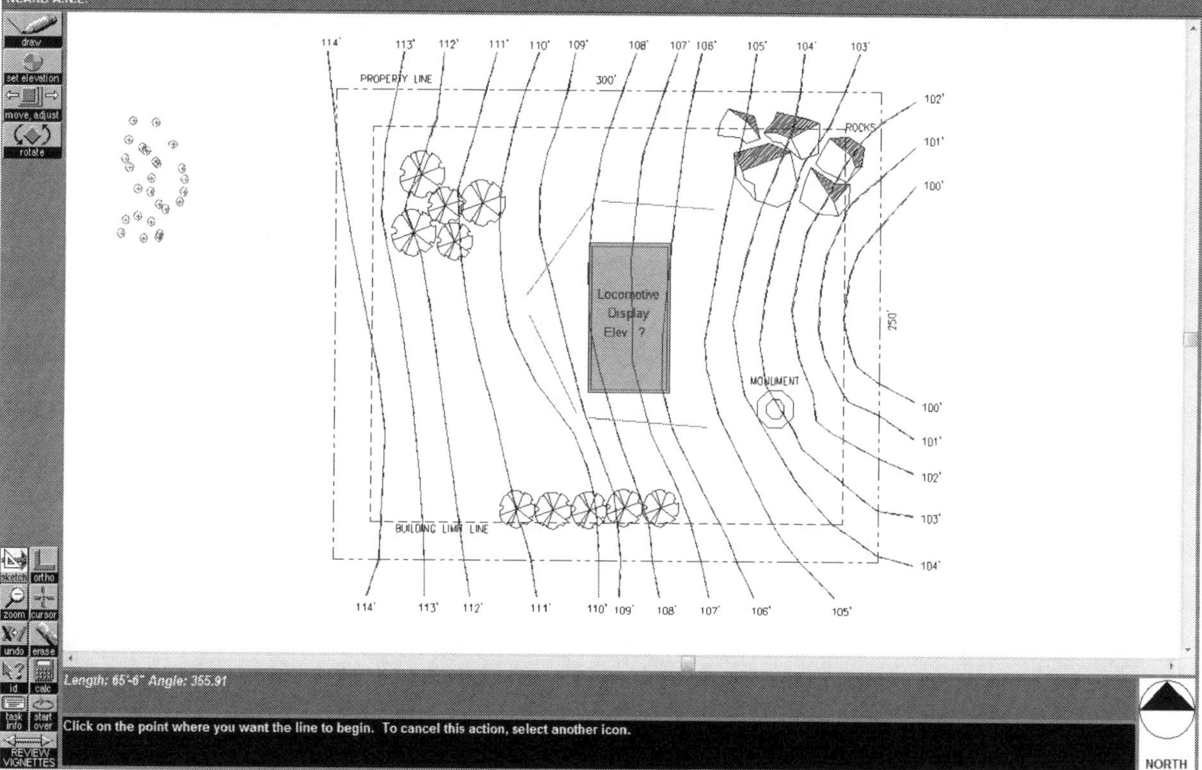

Figure 4.3 Use **Sketch > Line** to draw the centerlines of the two swales.

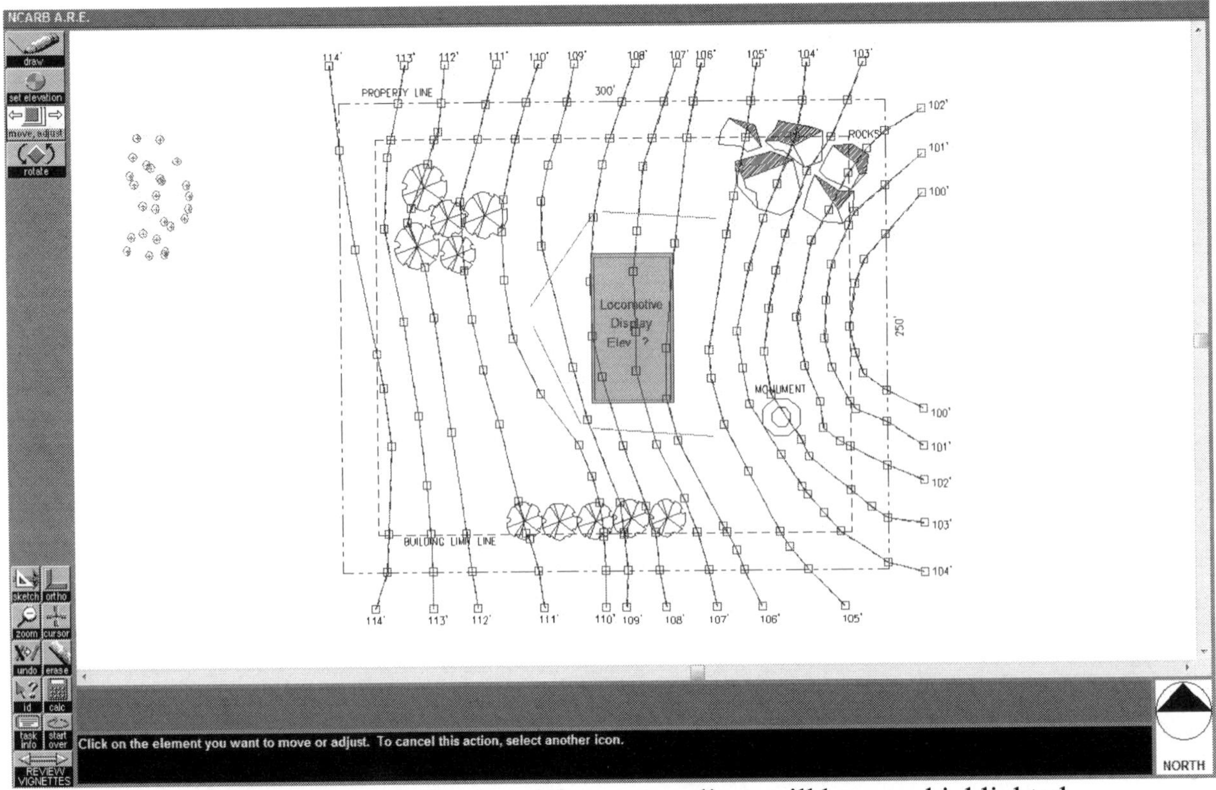

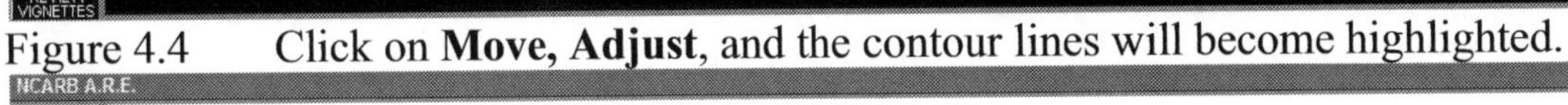
Figure 4.4 Click on **Move, Adjust**, and the contour lines will become highlighted.

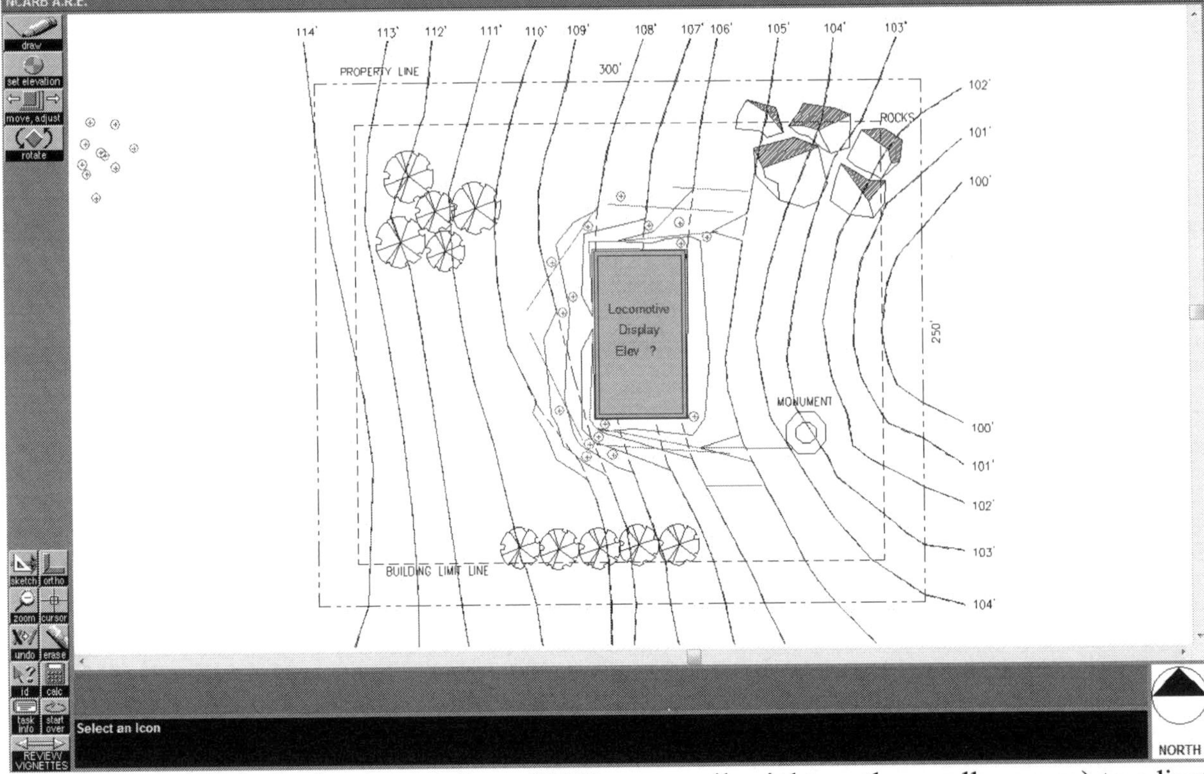

Figure 4.5 Click on the contour lines (NOT necessarily right on the small squares) to adjust them to form the two swales.

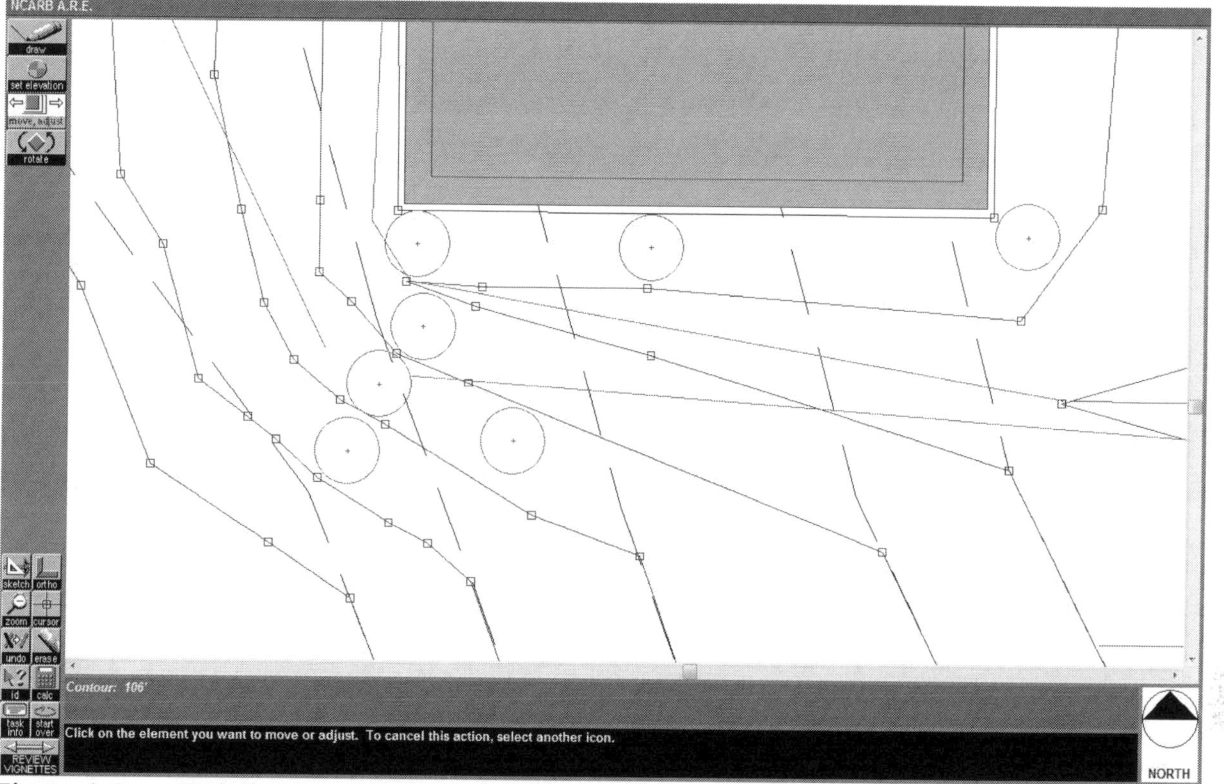

Figure 4.6 Use the pre-drawn 5'-diameter circles to check and make sure the horizontal distance between the adjacent contour lines is larger than 5'.

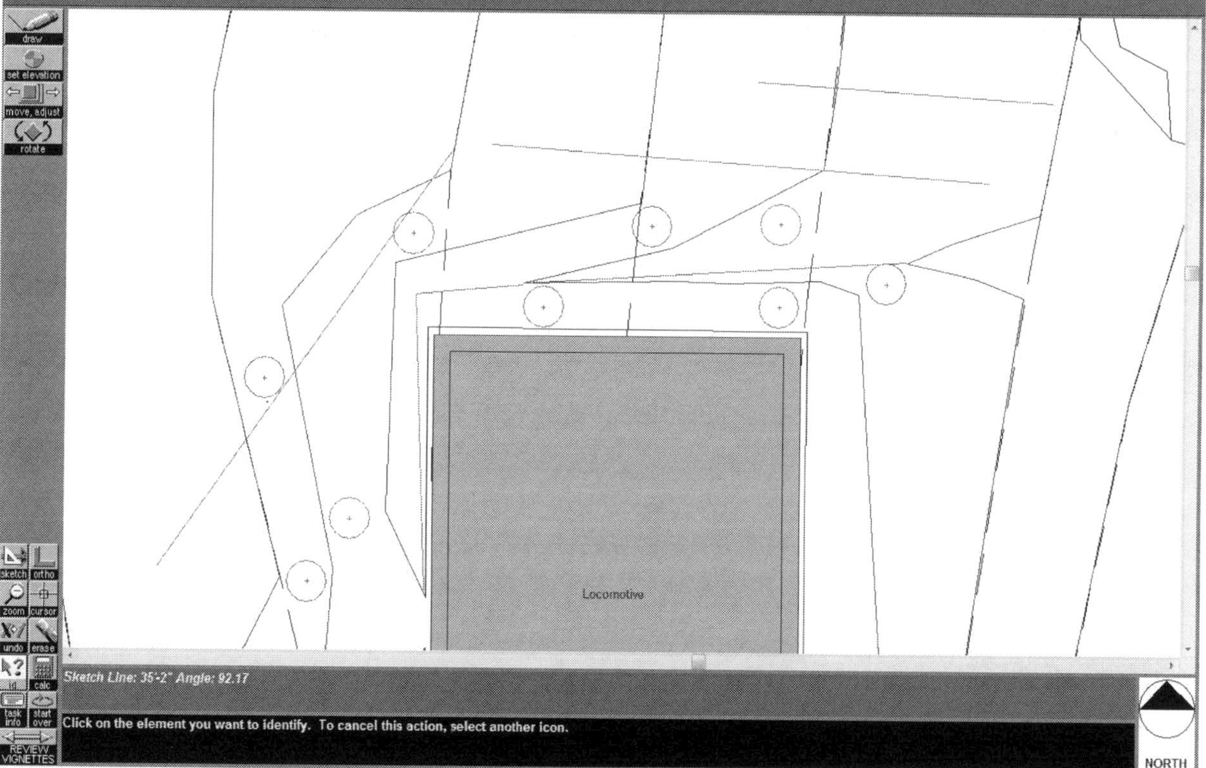

Figure 4.7 Use sketch line and id tool to check and make sure the horizontal distance between the adjacent contour lines is less than 50'.

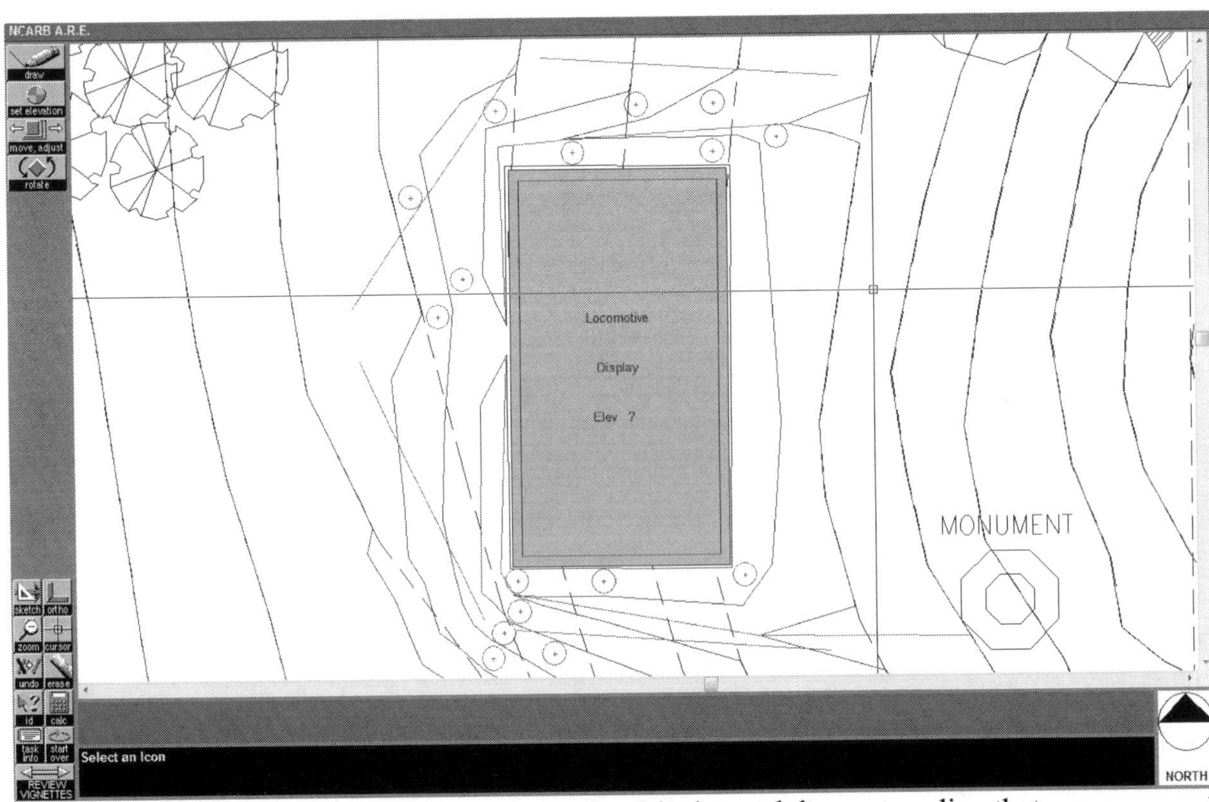

Figure 4.8 The slope between the Locomotive Display and the contour line that wraps around the Locomotive Display is not subject to the maximum slope requirement.

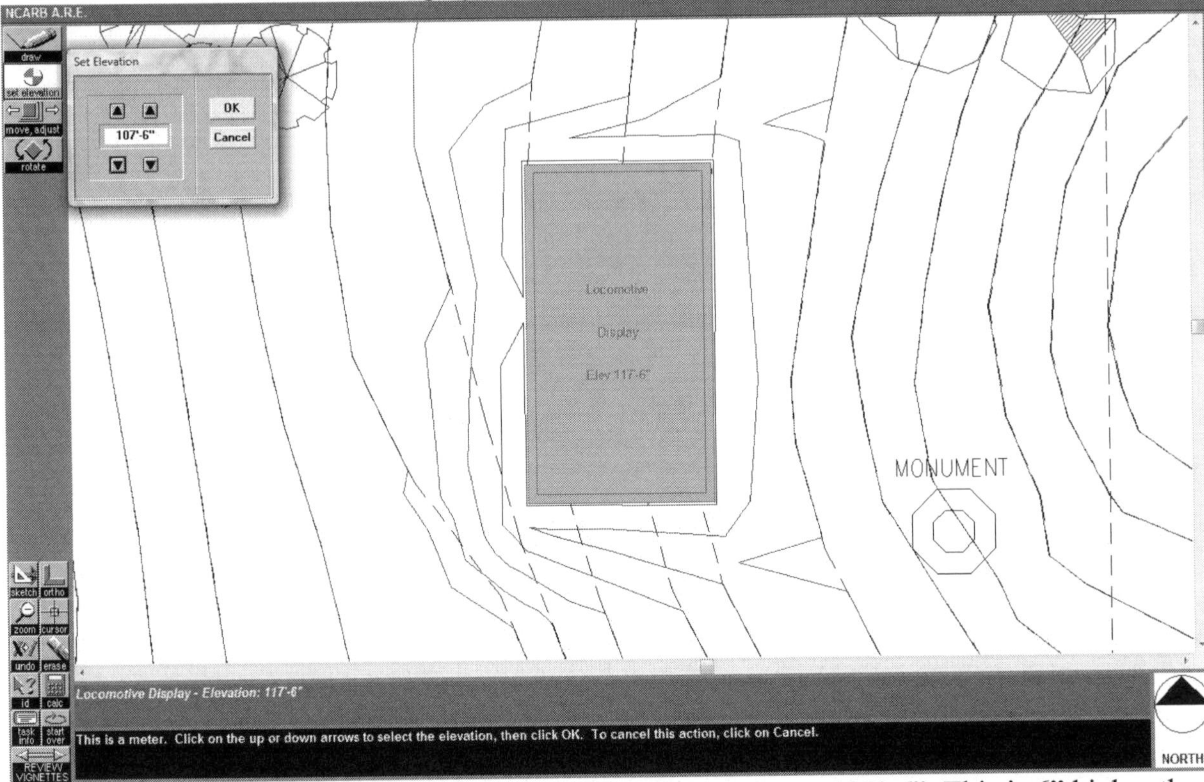

Figure 4.9 Use the **up and down arrow** to set the elevation to 107'-6". This is 6" higher than the 107' contour line that wraps around the Locomotive Display.

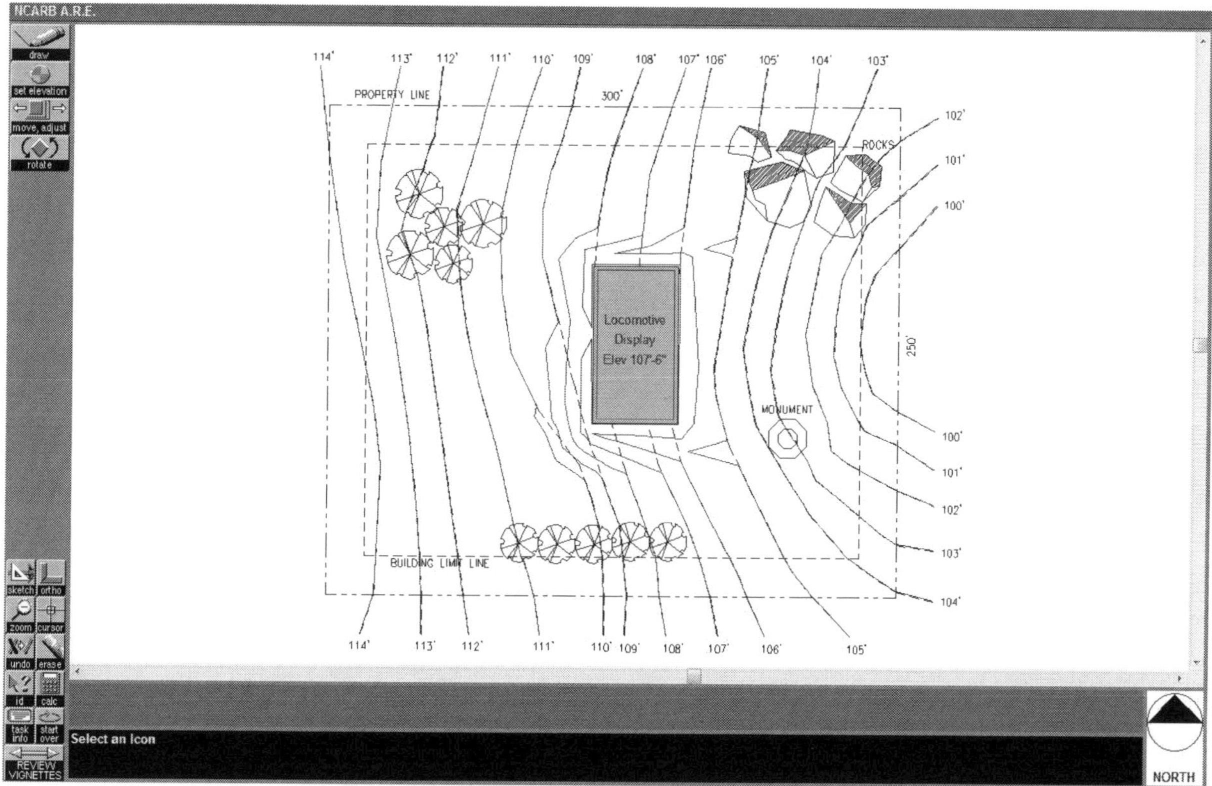

Figure 4.10 Click on the **Sketch > Hide sketch elements**, and the sketch lines and circles will disappear. This is your final solution.

8) Click on the **Sketch > Hide sketch elements**, and the sketch lines and circles will disappear. This is your final solution (Figure 4.10).

2. Items that you need to pay special attention to

Several items that you need to pay special attention to:

1) This vignette seems to be very simple, but you still need to check all the program requirements and make sure you comply with all of them.
2) Select a proper contour line as the contour line that wraps around the display area to create a level area.
3) The direction of the water flow is opposite to the "arrow" formed by a swale's contour lines.
4) Use the pre-drawn 5'-diameter circles to check and make sure the horizontal distance between the adjacent contour lines is larger than 5' (less than 20% slope). Adjust the contour lines if necessary.
5) Use sketch line and id tool to check and make sure the horizontal distance between the adjacent contour lines is less than 50' (larger than 2% slope). Adjust the contour lines if necessary.
6) The slope between the Locomotive Display and the contour line that wraps around the Locomotive Display is not subject to the maximum slope requirement, so place the contour line as close to the Locomotive Display as possible. Adjust the contour lines if necessary.
7) Make sure the contours around the monument, the rock and the trees are NOT disturbed.

C. Mock Exam Solution: Site Design Vignette

1. A Step-by-step solution to the graphic vignette: Site Design Vignette

1) **Overall concept:**

- The Site Design vignette has a very complicated program. If you simply read the program several times, you may NOT cover all the program requirements when you layout the site design. It is also VERY hard to understand and summarize the program requirements by simply reading it multiple times.
- One effective way of digesting the program requirements is to do a simple **Bubble Diagram**: use the paper and pencil provided by the test center to draw a simple **bubble diagram** to show the relationships between the site elements. This is the HARDEST part of the vignette.

  *Note: A **bubble diagram** is the MOST important tool for passing Site Design vignette.*

- The bubble diagram does NOT have to be pretty and does NOT have to be to scale. The key is to do the bubble diagram quickly and turn the program requirements into a graphic in the shortest time. The bubble diagram is for YOU. As long as YOU can read it, it is good enough. Architects love graphic thinking. A bubble diagram can be a great tool.
- A **bubble diagram** can be a few simple hand-drawn circles: A circle with a label such as OT represents the Office Tower; R represents the Restaurant; PP represents the Pedestrian Plaza; WW represents a walkway; an arrow between two circle represents direct access required between the two elements; an arrow with a number above it represents a clearance distance between the two elements; a line connecting two elements indicates that they need to be placed close to each other; an eye symbol represents an element should be visible from a certain location (Figure 4.11), etc.

2) The program requires us to provide only one 24 ft wide curb cut located no closer than 110 ft from the intersection of the centerlines of the two existing public streets. Use **Sketch > Circle** to draw a circle with a 110-ft radius and with its center at the intersection of the centerlines of the two existing public streets; the 24 ft wide curb cut must be outside of this sketch circle (Figure 4.12).

3) Provide a 30 ft setback from the Pond for all construction or built improvements. Use **Zoom** to zoom into the pond area. Use **Sketch > Circle** to draw a number of circles with 15 ft radii defining the 15 ft setback from the pond (Figure 4.13).

   Note:

   - *The radius of the circle displays on the lower-left-hand corner of the screen when drawn. After you draw the first circle, the radius for rest of the circles will stay at 15 ft, and you just need to click on the screen to place them.*

4) Locate the 5-story, 60 ft high Office Tower close to the Pond, and the main entrance shall be visible from Bentley Avenue. Use **Zoom** to zoom out. Use **Draw** > **Office Tower** to draw the office tower near the pond with its main entrance facing Bentley Avenue (Figure 4.14).

5) Use **Draw** > **Restaurant** to draw the restaurant (Figure 4.15).

6) The main entrance of the Restaurant shall receive the noonday summer sun: Assume a 45° solar altitude angle. This means the main entrance of the Restaurant shall face south and need to avoid the shadow of the 60 ft high Office tower and the existing 35ft high shopping center. Click on **Rotate**, click on the restaurant, and click on **Rotate** again to rotate the restaurant. Use **Move, Adjust** to move it up to make room for the Pedestrian Plaza (Figure 4.16).

   Note:
   - *Pay attention to the legend on the lower-right hand corner of the screen. The main entrance is marked with a* ***solid-filled*** *triangle, and the service entrance is marked with a* ***regular*** *triangle.*
   - *You can use* ***Sketch*** *>* ***Rectangle*** *to draw a rectangle with a 60 ft side to simulate the shadow of the Office Tower with a 45° solar altitude angle.*

7) Buildings must be separated by a minimum of 20 ft. Use **Sketch** > **Line** to check and make sure the distance between the Office Tower and the Restaurant is larger than 20 feet. Check to make sure the distance between the Office Tower and the Existing Shopping Center is larger than 20 feet also.

   Note:
   - *You can use* ***Sketch*** *>* ***Circle*** *to draw a circle with a 20-ft radius and with its center at a corner of the Office Tower to define the 20 ft minimum separation between the Office Tower and the Restaurant.*

8) Click on **Ortho** to turn on the Ortho mode. Use **Zoom** to zoom in. Use **Draw** > **Pedestrian Plaza** to draw the Pedestrian Plaza (Figure 4.17).

9) Click on **id?** then click on the Pedestrian Plaza. The square footage (8,786sf) of the Pedestrian Plaza appears on the lower-left-hand corner of the screen. The actual square footage of the Pedestrian Plaza is too small. Use **Move, Adjust** to adjust the square footage to as close to 9,000 sf, the required square footage, as possible. Use **Zoom** to zoom in if necessary. (Figure 4.18). We are able to adjust the square footage of the Pedestrian Plaza to 8,978 sf. This is very close to 9,000 sf.

   Note: *In any case, the Pedestrian Plaza should NOT be 10% larger or smaller than the required size.*

10) Locate the universally accessible parking spaces within 100 ft of the main entrance of the Office Tower. Use **Sketch** > **Circle** to draw a circle with a 100-ft radius and with its center at the main entrance of the Office Tower (Figure 4.19).

11) 3 universally accessible (12 ft x 18 ft) parking spaces are required. Click on **Draw** > **Handicap Spaces**, a dialogue box will pop up. Click on the **down arrow** to set the number of Handicap Spaces to 3 (Figure 4.20).

12) Click on **OK**, and then Click on **Ortho** to turn on the Ortho mode. The Handicap Spaces are drawn as 3-point rectangle. Set the over widths of the Handicap Spaces to 36' (12'x3=36'), and the depth of the Handicap Spaces to 18'. The dimensions of the Handicap Spaces appear on the lower-left-hand corner of the screen (Figure 4.21).

    Note: *You should ONLY use* ***Draw > Handicap Spaces*** *to draw the handicap spaces. Do NOT use* ***Draw > Standard Spaces*** *to draw the handicap spaces.*

13) Drives, traffic aisles, and parking spaces shall be no closer than 5 ft to a building. Use **Draw** > **Sketch Rectangle** to draw a 5' wide sketch rectangles to set the 5-ft clearance for the Office Tower (Figure 4.22).

14) Use **Rotate** to rotate the 3 Handicap Spaces. Use **Move, Adjust** to move them closer to the Office Tower (Figure 4.23).

15) The intersection of the access drive with the street must be perpendicular to the street for at least the first 20 ft of the drive. Use **Draw** > **Driveway** to draw a drive from Bentley Avenue, outside of the utility easement, and at least 5' from the Restaurant (Figure 4.24).

    Note: *Use Draw >* ***Sketch Rectangle*** *to draw a 24' wide sketch rectangles to assist you to locate the driveway if necessary.*

16) Use **Draw** > **Driveway** to draw a service drive to the service entrance of the Restaurant (Figure 4.25).

17) 30 standard (9 ft x 18 ft) parking spaces are required. Use **Draw** > **Standard Spaces** to draw 30 standard spaces. Pay attention to the following requirements (Figure 4.26):
    - Drive-through circulation is required.
    - Dead-end parking is prohibited.
    - Parking along the service drive is prohibited.

    Note:
    - *Use Draw >* ***Sketch Rectangle*** *to draw a 24' wide sketch rectangles to assist you to locate the driveways and parking spaces if necessary.*
    - *If a space is too narrow for placing a driveway, you can use move group to move the other elements to make room for the driveway.*
    - *We also adjusted the Pedestrian Plaza to make room for the parking spaces.*

18) Use **Draw > Driveway** to draw the remaining driveways, adjust the driveway to make sure they join properly. You may need to move or adjust the Restaurant, the Pedestrian Plaza, and the Office Tower again to make sure all the elements fit together (Figure 4.27).

Note:

- *The key to draw, align or join different sections of the driveway is to align the centerlines of the driveways.*
- *Use Move, Adjust to adjust the length of the driveways if necessary.*
- *Sometimes it may be easier to erase a driveway, and then redraw it instead of trying to adjust it.*
- *Make sure to maintain the square footage of the Pedestrian Plaza per the program if you adjust it.*

19) Use **Draw > Walkway** to draw a sidewalk from the Handicap Spaces to the Pedestrian Plaza, and a sidewalk from the Pedestrian Plaza to the Public Walk along Bentley Avenue (Figure 4.28).

20) The service entrance is NOT facing the Pedestrian Plaza. It is already blocked by the Restaurant itself from the Pedestrian Plaza (Figure 4.29).

Note: *The view of the service entrance (NOT the entire service entrance and loading area) on the Restaurant shall be blocked from the Pedestrian Plaza by buildings and/or trees.*

21) The Pedestrian Plaza shall be blocked from the prevailing winter winds by buildings and/or trees. Pay attention to the *arrow marked with "winds"* on the upper-right hand corner of the screen. Use **Sketch > Line** to draw a few sketch lines parallel to the direction of wind to assist you to place coniferous trees to block the Pedestrian Plaza from the prevailing winter winds (Figure 4.30).

Note: *The sketch lines parallel to the direction of winds are very valuable tool for you to layout the coniferous trees.*

22) Use **Draw > Coniferous Tree** to draw several coniferous trees to block the winds (Figure 4.31).

Note:

- *Do NOT overlap trees.*
- *Make sure the main entrance of the Office Tower is visible from Bentley Avenue when adding trees.*
- *Pay attention to the elevation of trees: Deciduous trees will allow views under the tree crown; coniferous trees have a triangle shape, and may still allow view or winds to pass through if you simply place them side by side.*
- *Do NOT draw more trees than necessary because you may accidentally block some other elements or violate another program requirement.*

23) Use **Sketch > Hide Sketch Elements** to hide sketch elements (Figure 4.32).

24) Click on **Check**, all the trees that overlap with sidewalk, parking space or other elements will be highlighted (Figure 4.33). We have 5 highlighted trees, but only 3 of them need to be cut. This is less than the 6 cut tress as allowed by the program.

Note:

- *If a tree overlaps a parking space or sidewalk, it may not need to be cut. By reviewing the elevation of the trees provided by the NCARB program, you notice that a small portion of the crown of a deciduous tree can overlap a parking space or sidewalk. On the other hand, if the crown of a coniferous tree overlaps a parking space or sidewalk, the tree will be cut.*
- *We have to block the Pedestrian Plaza from the winter winds, but we also need to make sure the main entrance of the Office Tower is visible from Bentley Street. This is a way to show architecture is a balancing act.*

25) After finishing your solution, re-read the program again carefully to double check and make sure you cover every requirement.

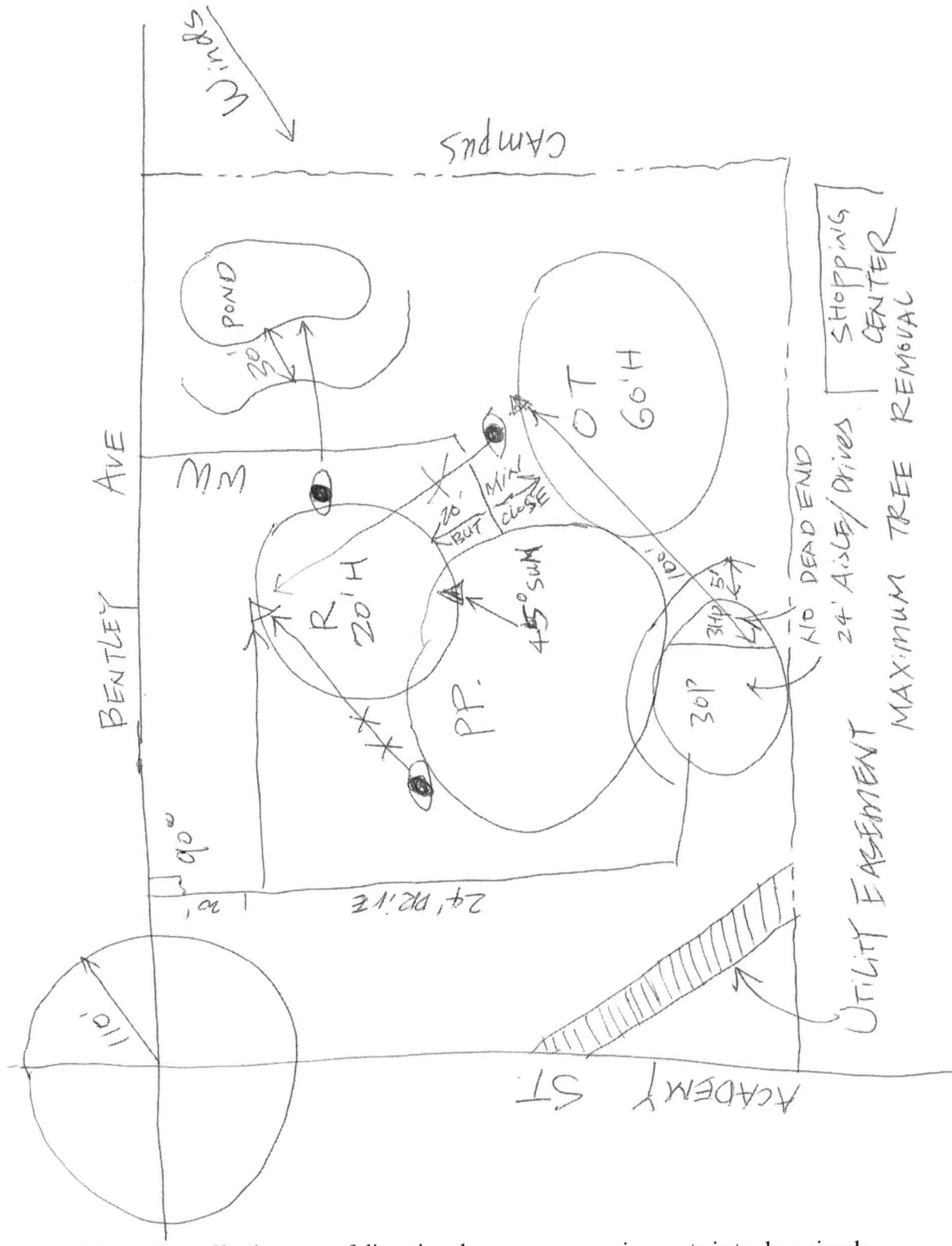

Figure 4.11 One effective way of digesting the program requirements is to do a simple **Bubble Diagram**.

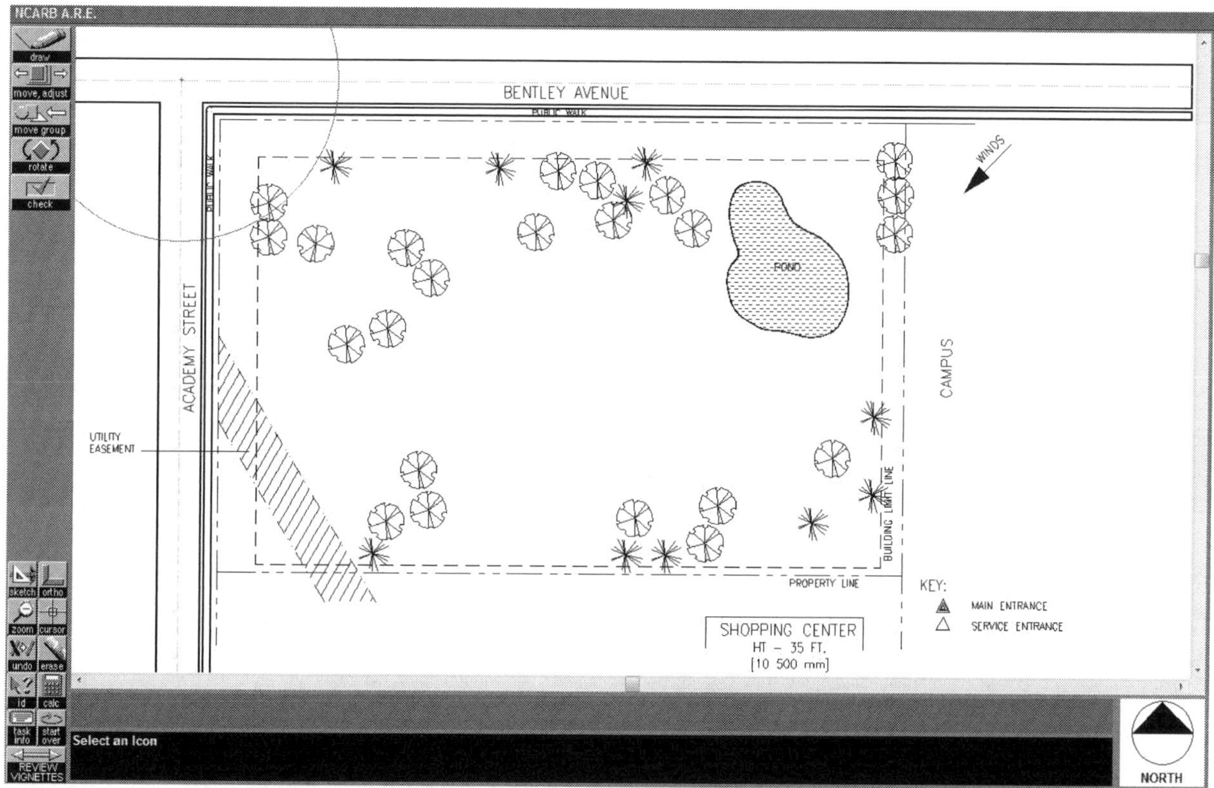

Figure 4.12 Use **Sketch > Circle** to draw a circle with its center at the intersection of the centerlines of the two existing public streets.

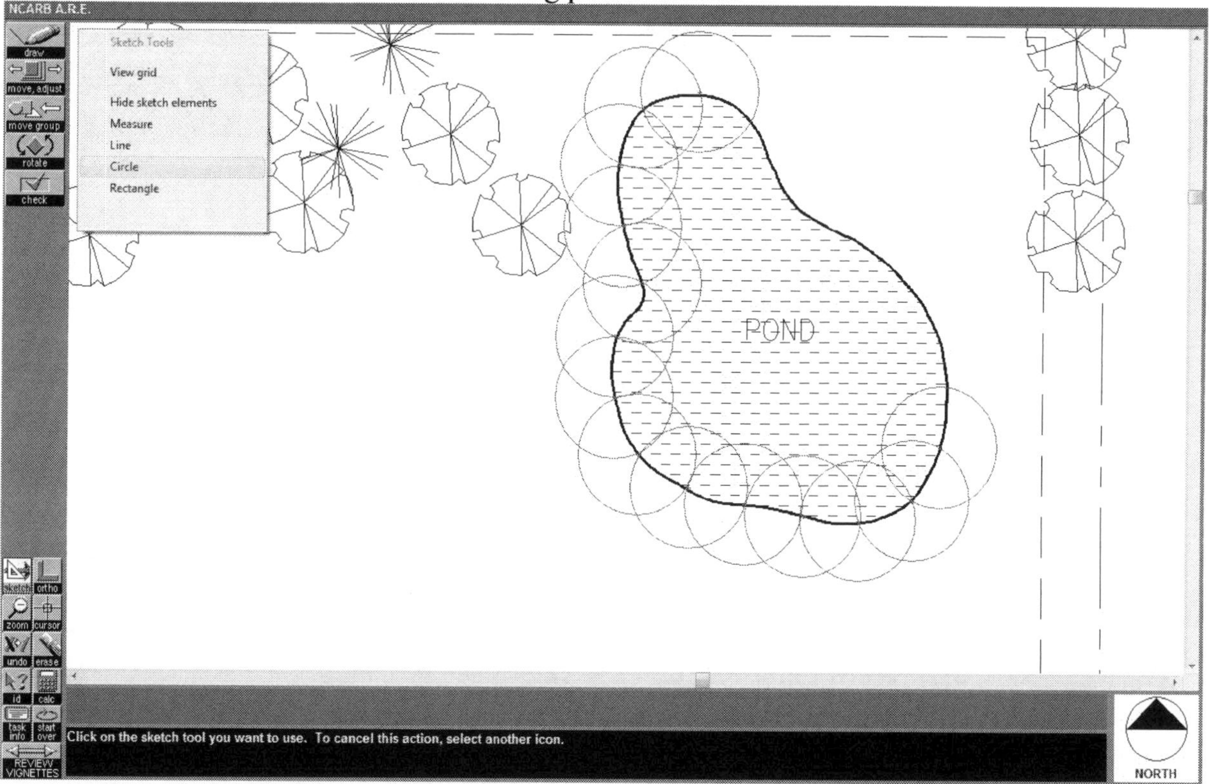

Figure 4.13 Use **Sketch > Circle** to draw a number of circles with 15ft radii defining the 15 ft setback from the pond.

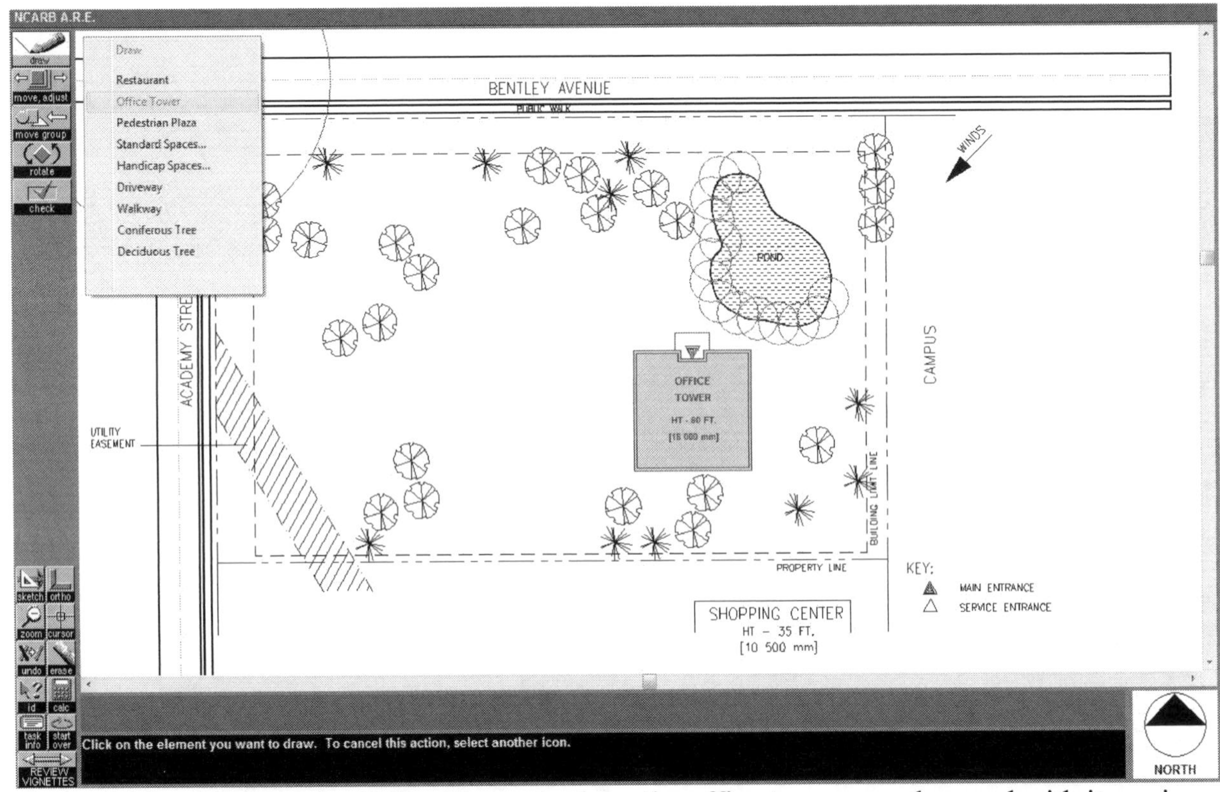

Figure 4.14 Use **Draw > Office Tower** to draw the office tower near the pond with its main entrance facing Bentley Avenue.

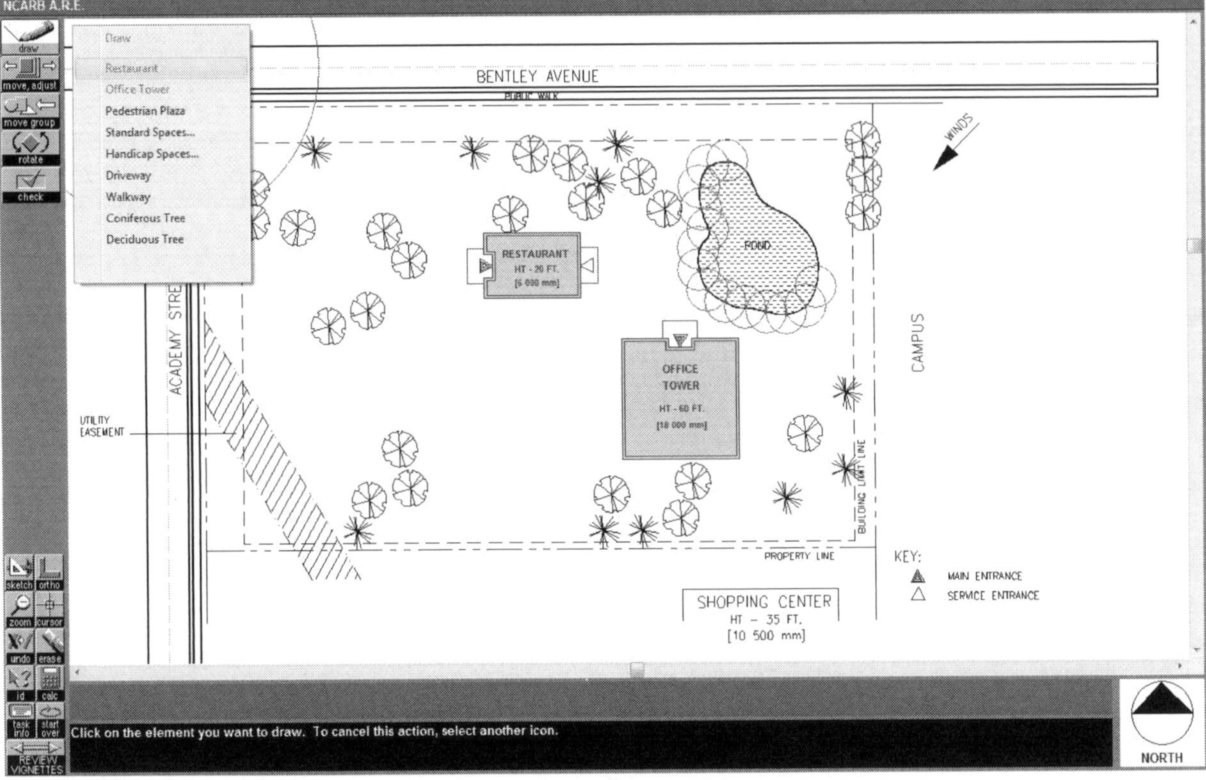

Figure 4.15 Use **Draw > Restaurant** to draw the restaurant.

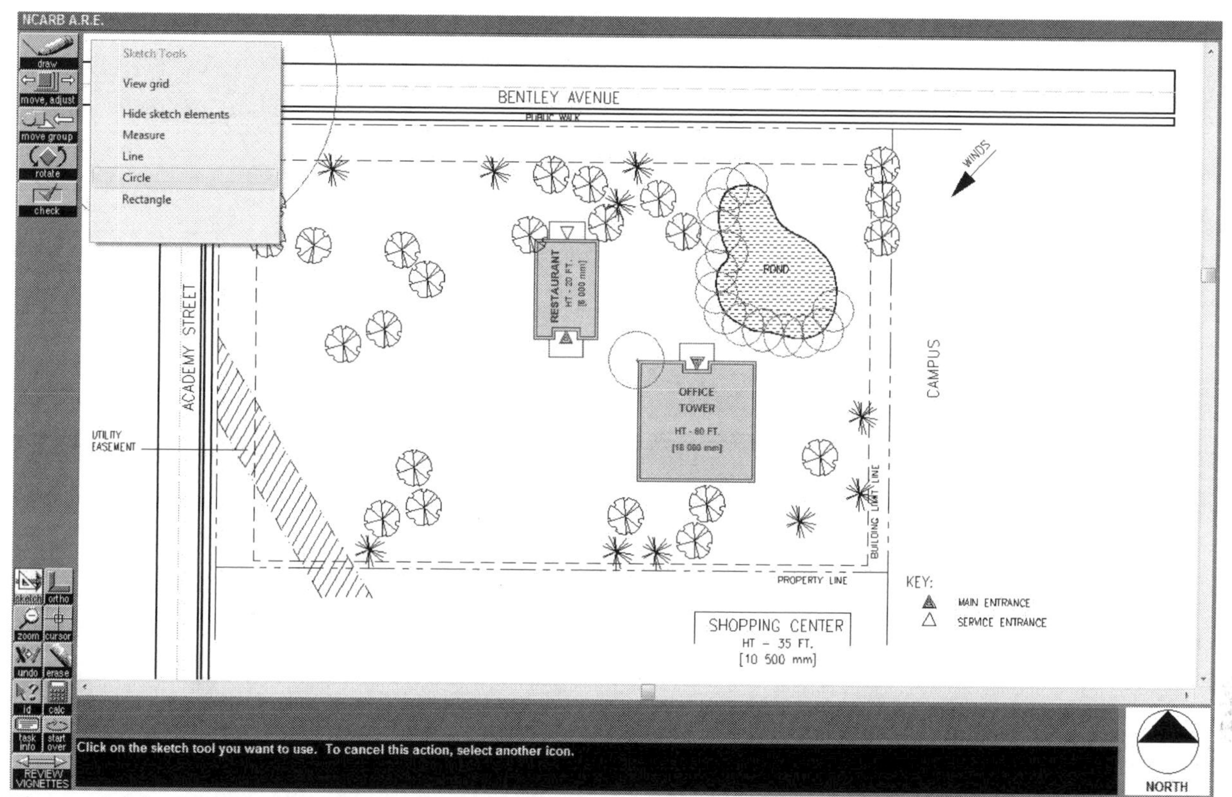

Figure 4.16 Click on **Rotate**, click on the restaurant, and click on **Rotate** again to rotate the restaurant.

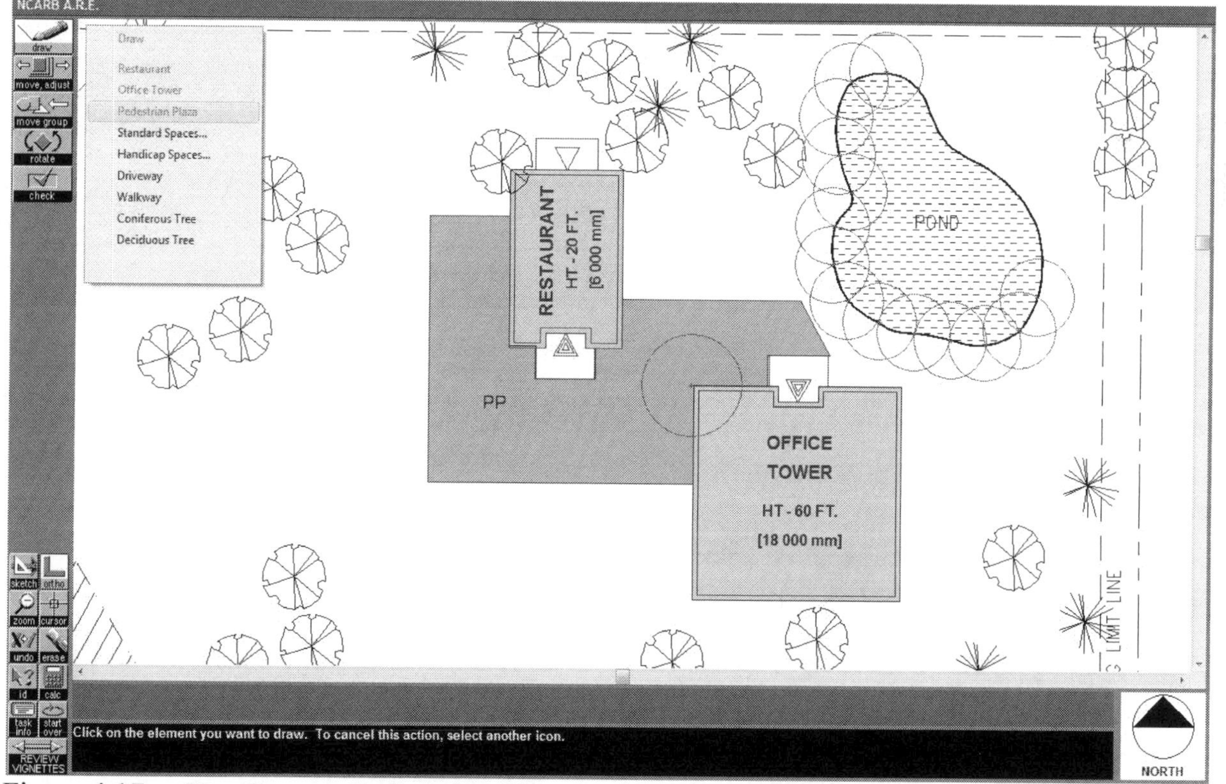

Figure 4.17 Use **Draw > Pedestrian Plaza** to draw the Pedestrian Plaza.

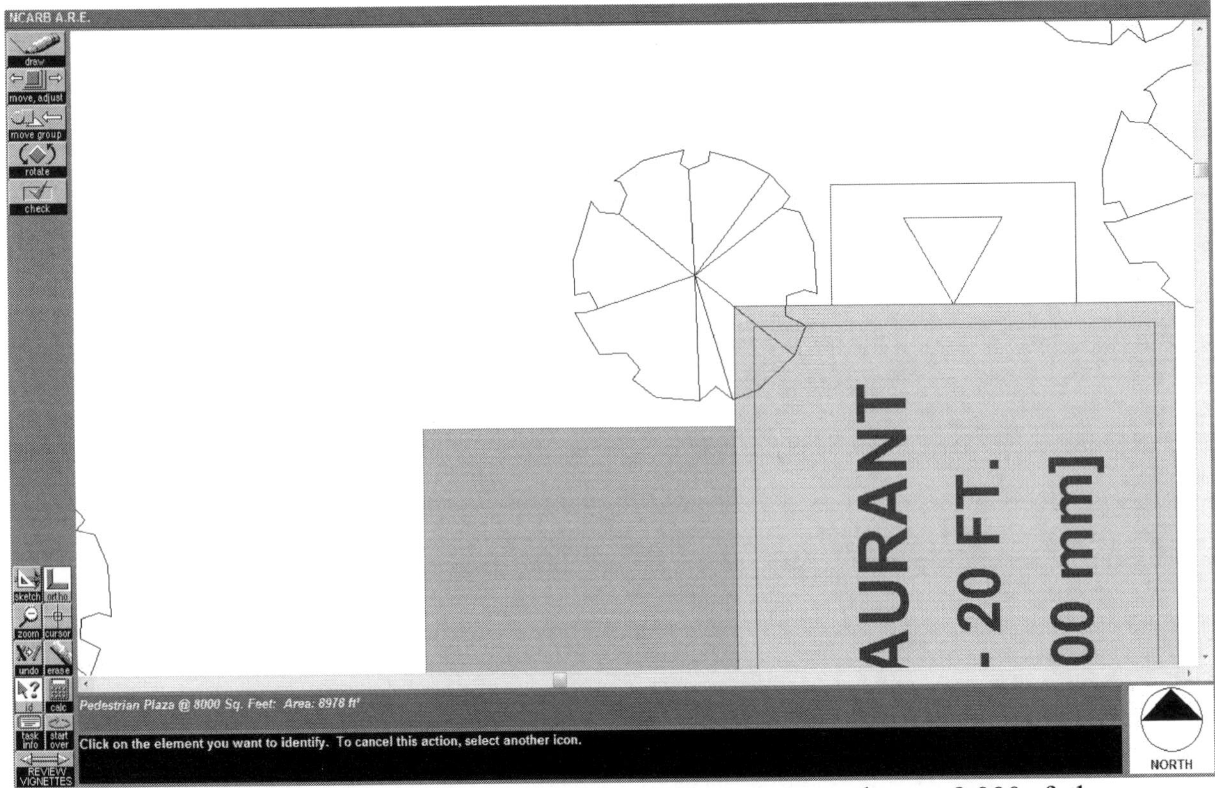

Figure 4.18 Use **Move, Adjust** to adjust the square footage to as close to 9,000 sf, the required square footage, as possible. Use **Zoom** to zoom in if necessary.

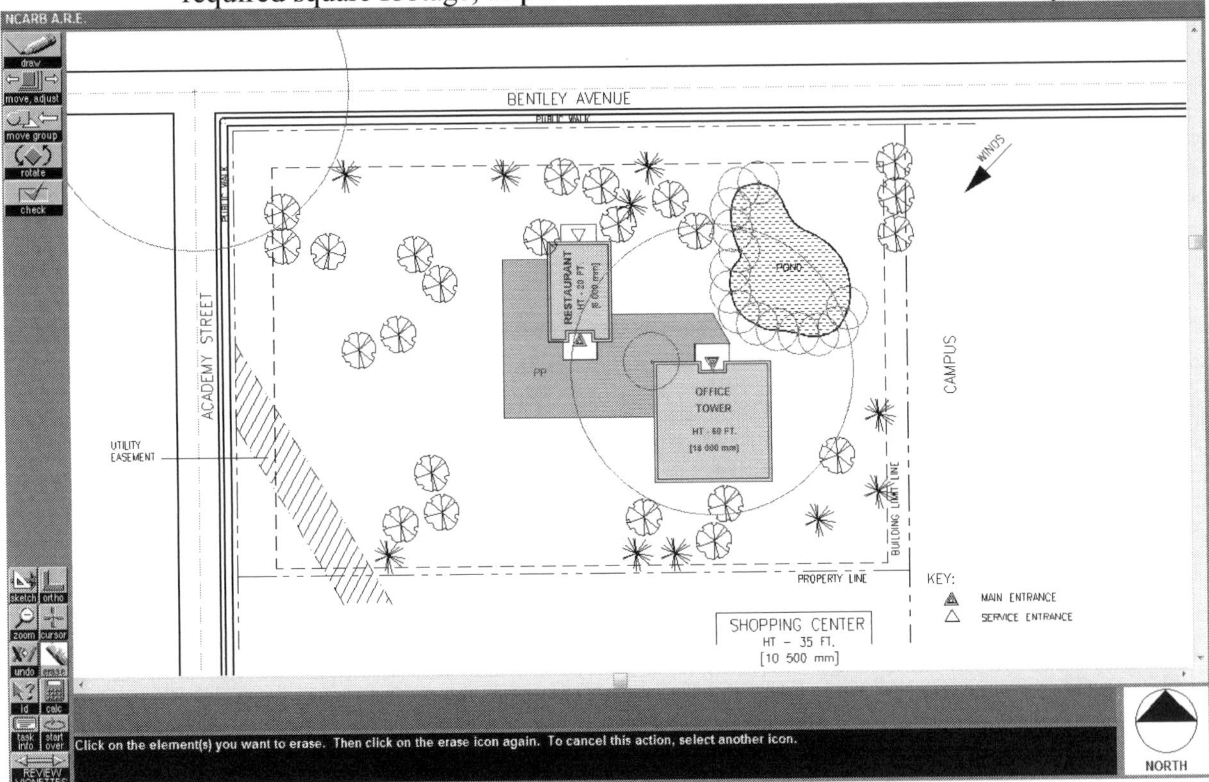

Figure 4.19 Use **Sketch > Circle** to draw a circle with a 100-ft radius and with its center at the main entrance of the Office Tower.

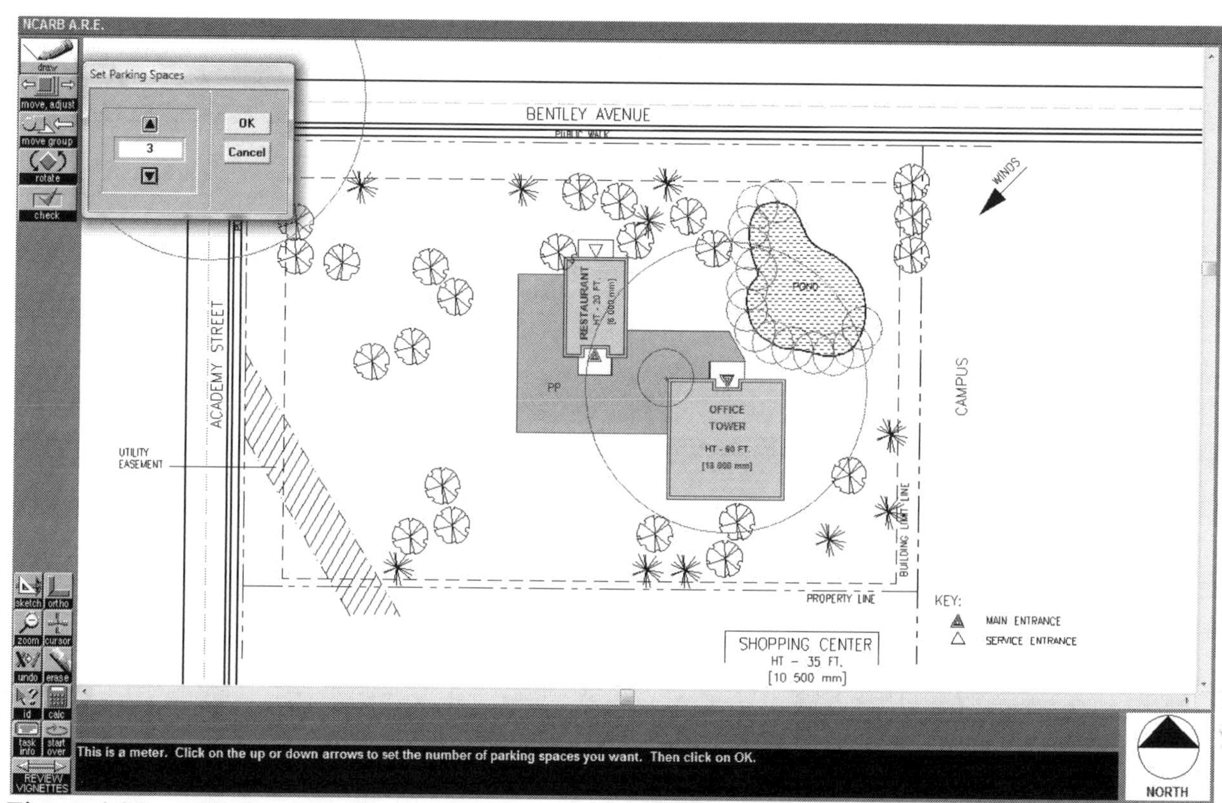

Figure 4.20 Click on the down arrow to set the number of Handicap Parking spaces to 3.

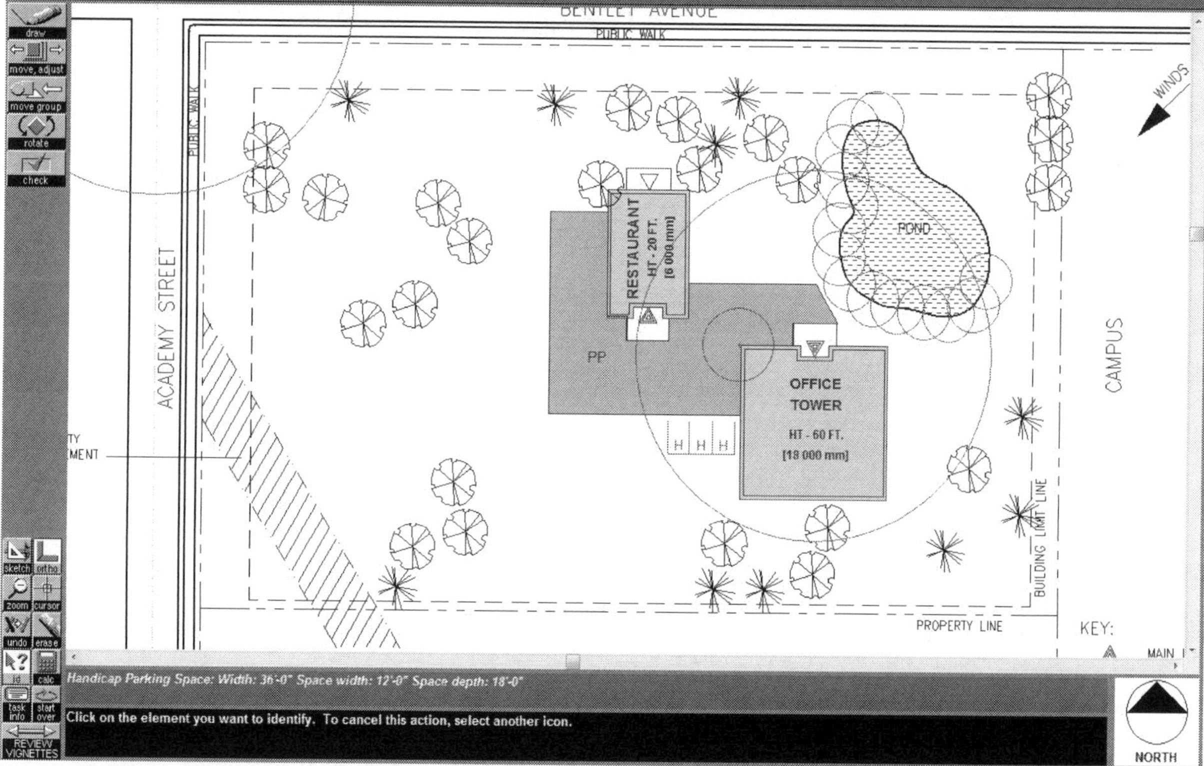

Figure 4.21 Set the over widths of the Handicap Spaces to 36' (12'x3=36'), and the depth of the Handicap Spaces to 18'.

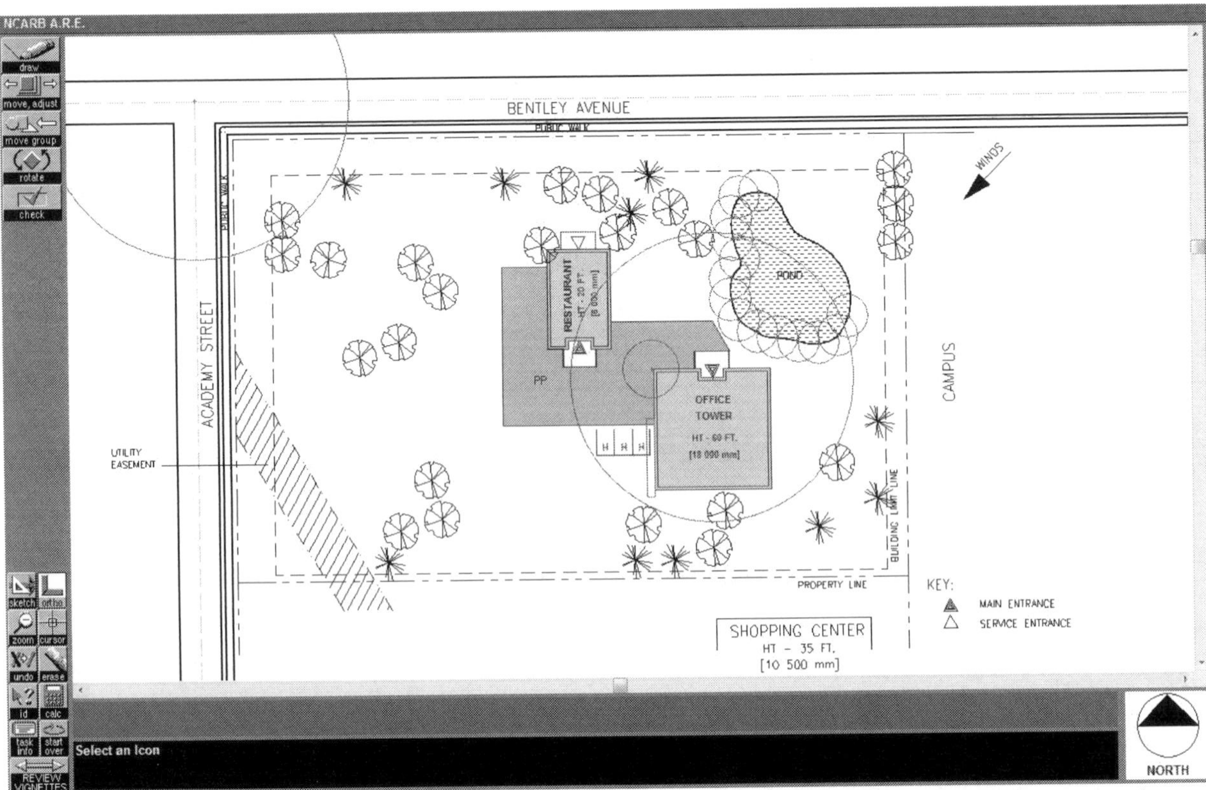

Figure 4.22 Use **Draw > Sketch Rectangle** to draw the 5' wide sketch rectangles to set the 5-ft clearance for the Office Tower.

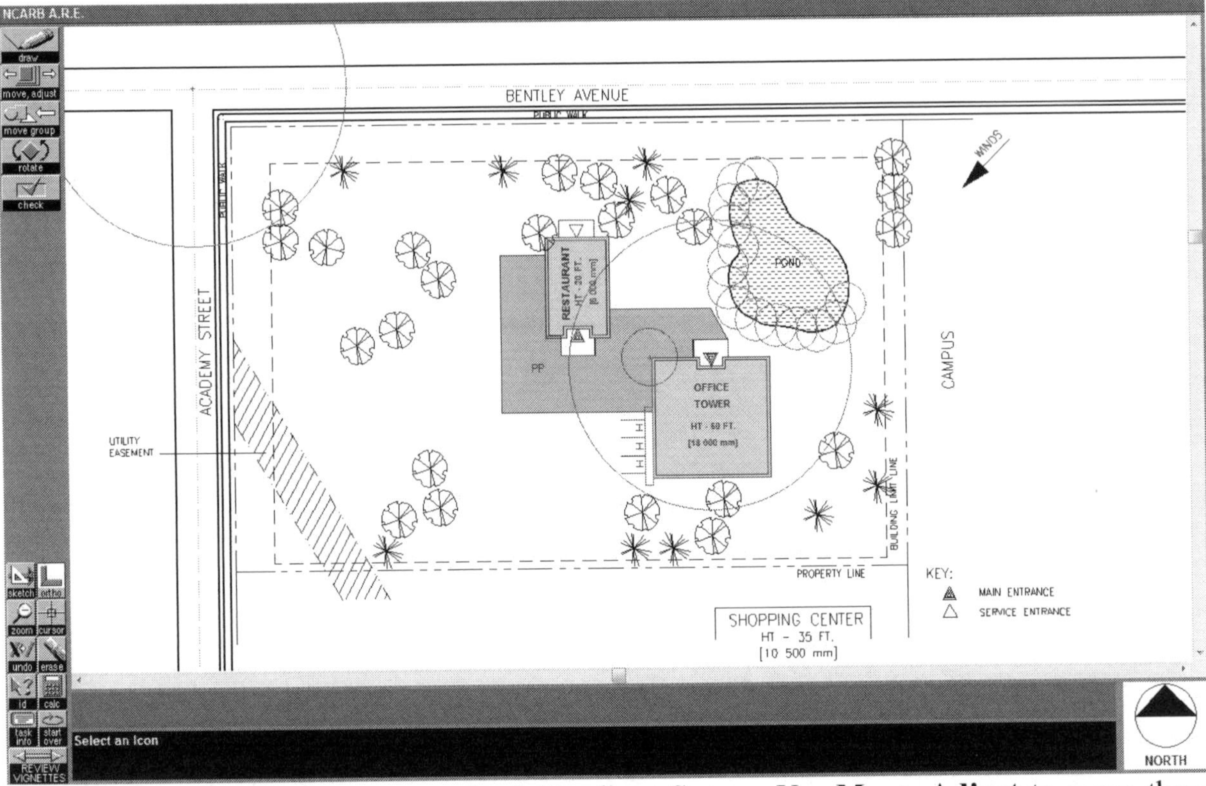

Figure 4.23 Use **Rotate** to rotate the 3 Handicap Spaces. Use **Move, Adjust** to move them closer to the Office Tower.

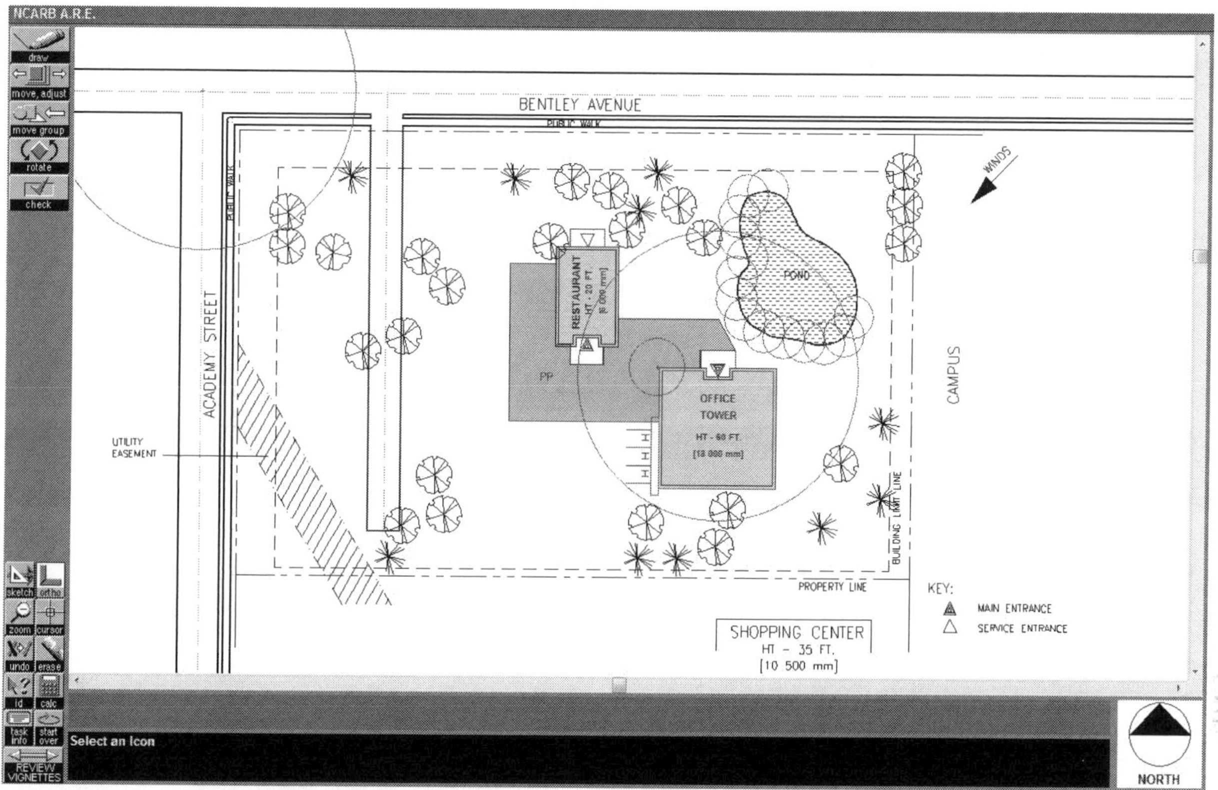

Figure 4.24 Use **Draw > Driveway** to draw a drive from Bentley Avenue, outside of the utility easement, and at least 5' from the Restaurant.

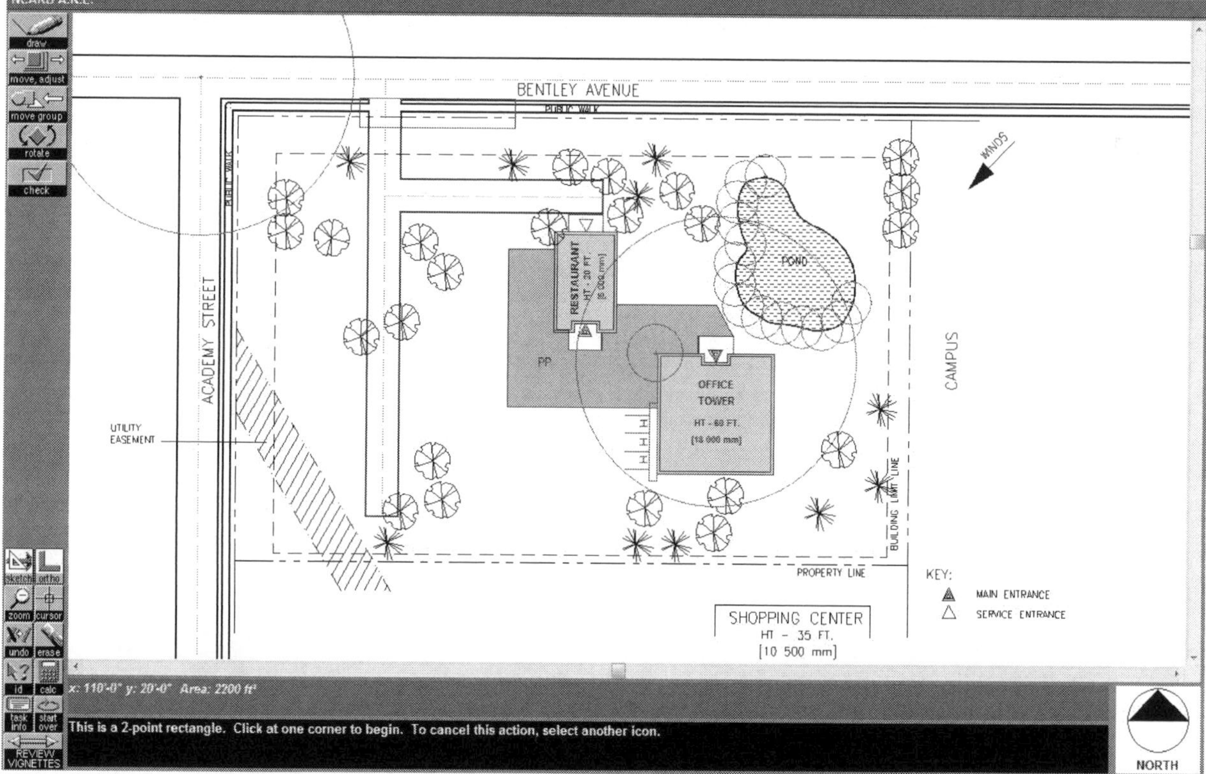

Figure 4.25 Use **Draw > Driveway** to draw a service drive to the service entrance of the Restaurant.

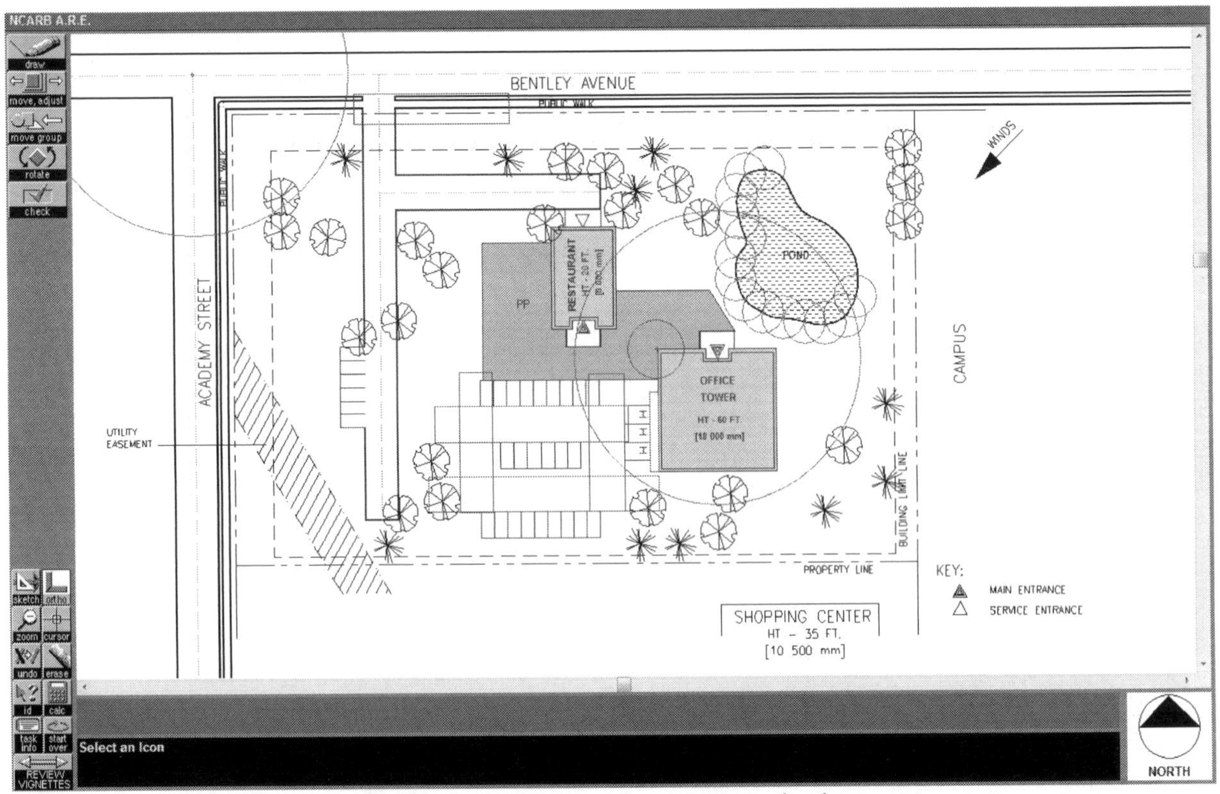

Figure 4.26 Use **Draw > Standard Spaces** to draw 30 standard spaces.

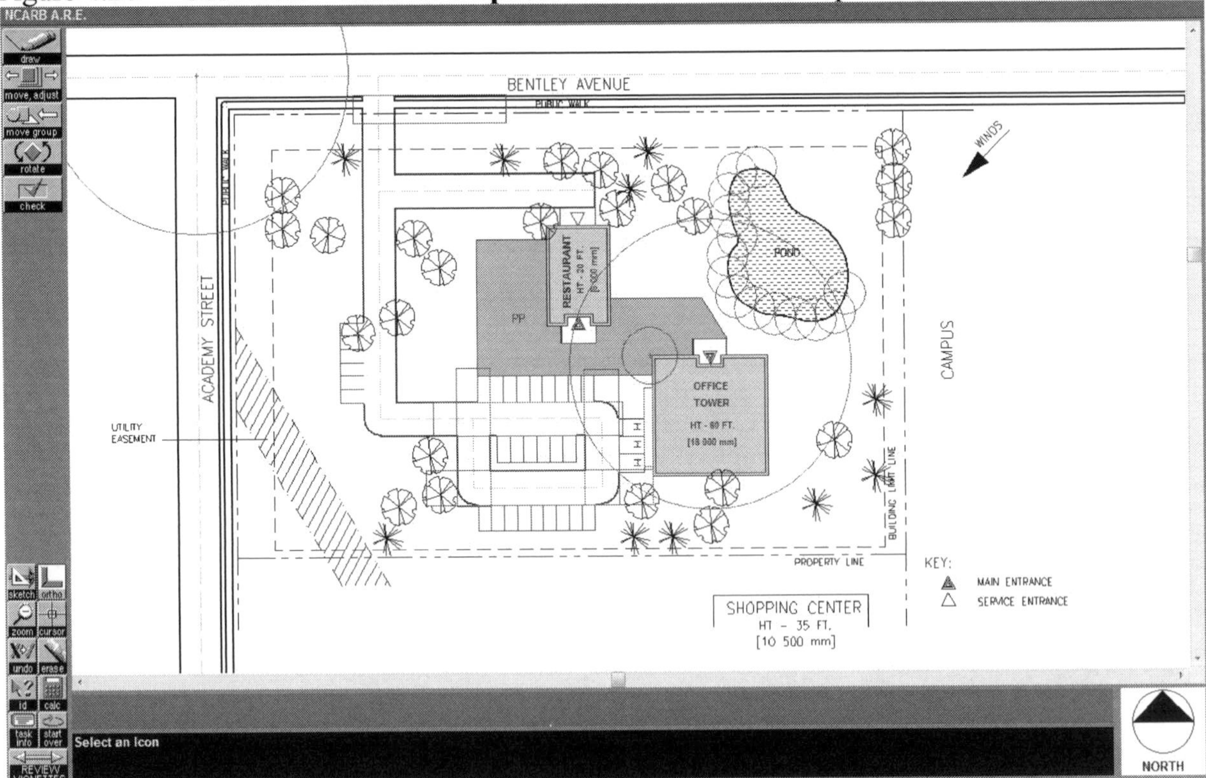

Figure 4.27 Use **Draw > Driveway** to draw the remaining driveways.

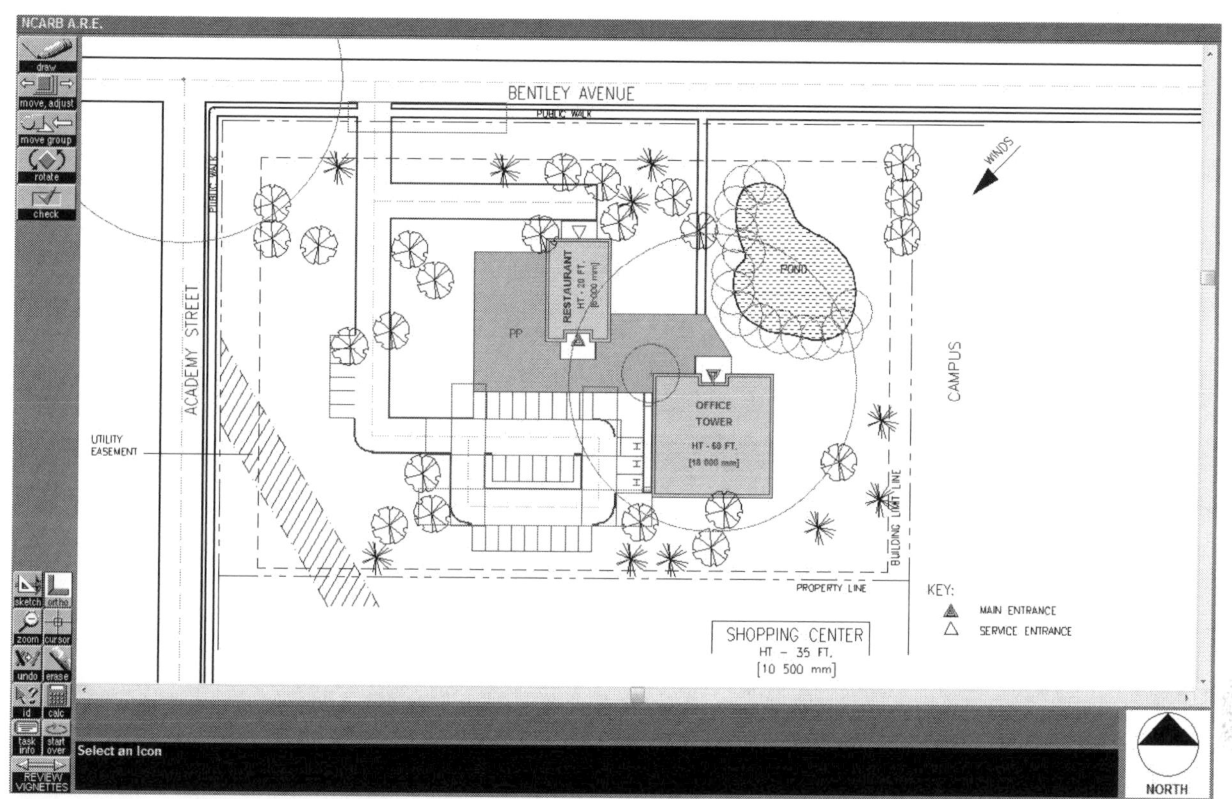

Figure 4.28 Use **Draw > Sidewalk** to draw sidewalks.

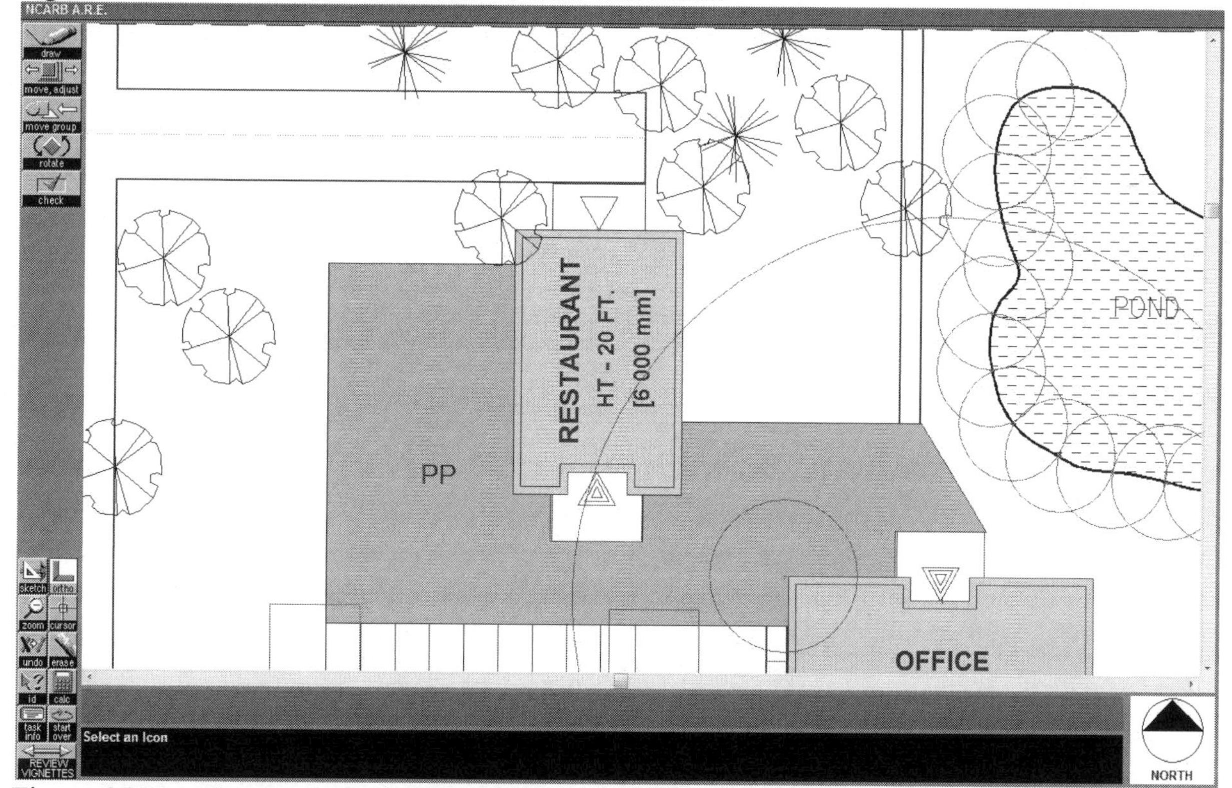

Figure 4.29 The service entrance is NOT facing the Pedestrian Plaza. It is already blocked by the Restaurant itself from the Pedestrian Plaza

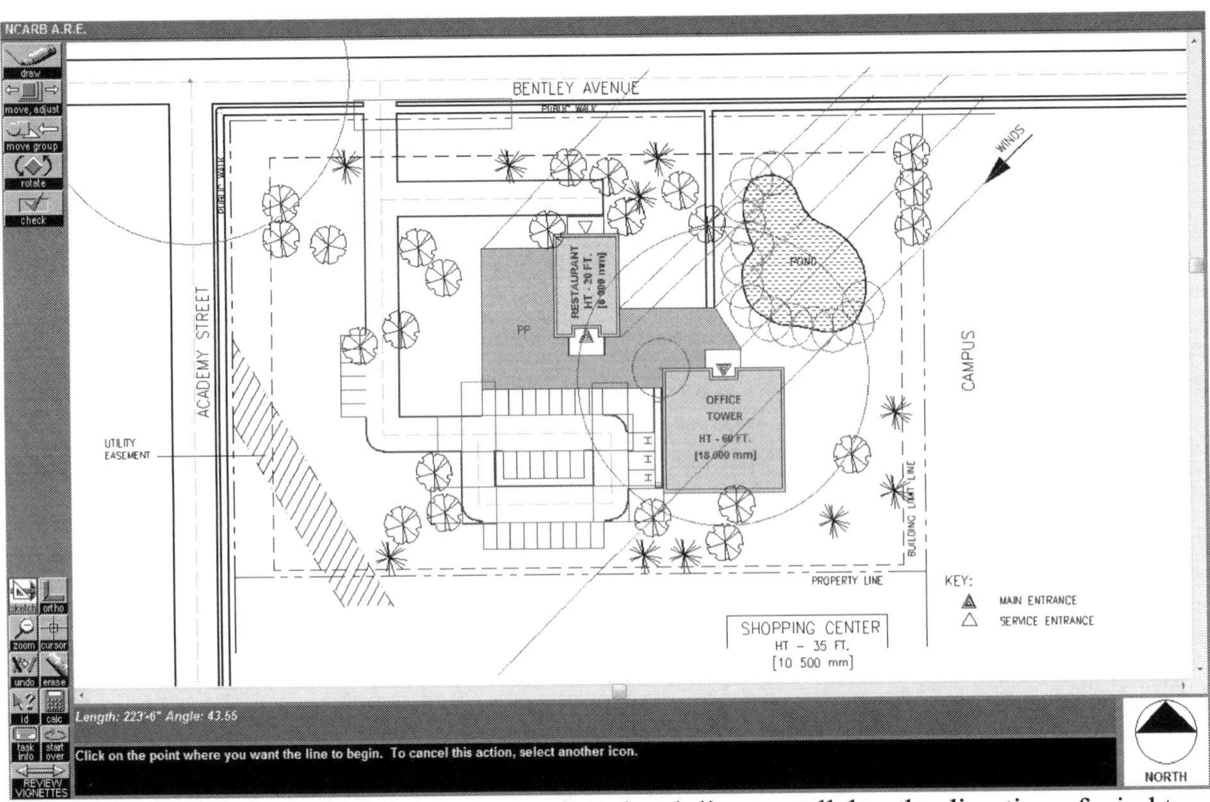

Figure 4.30 Use **Sketch > Line** to draw a few sketch lines parallel to the direction of wind to assist us to locate the coniferous trees.

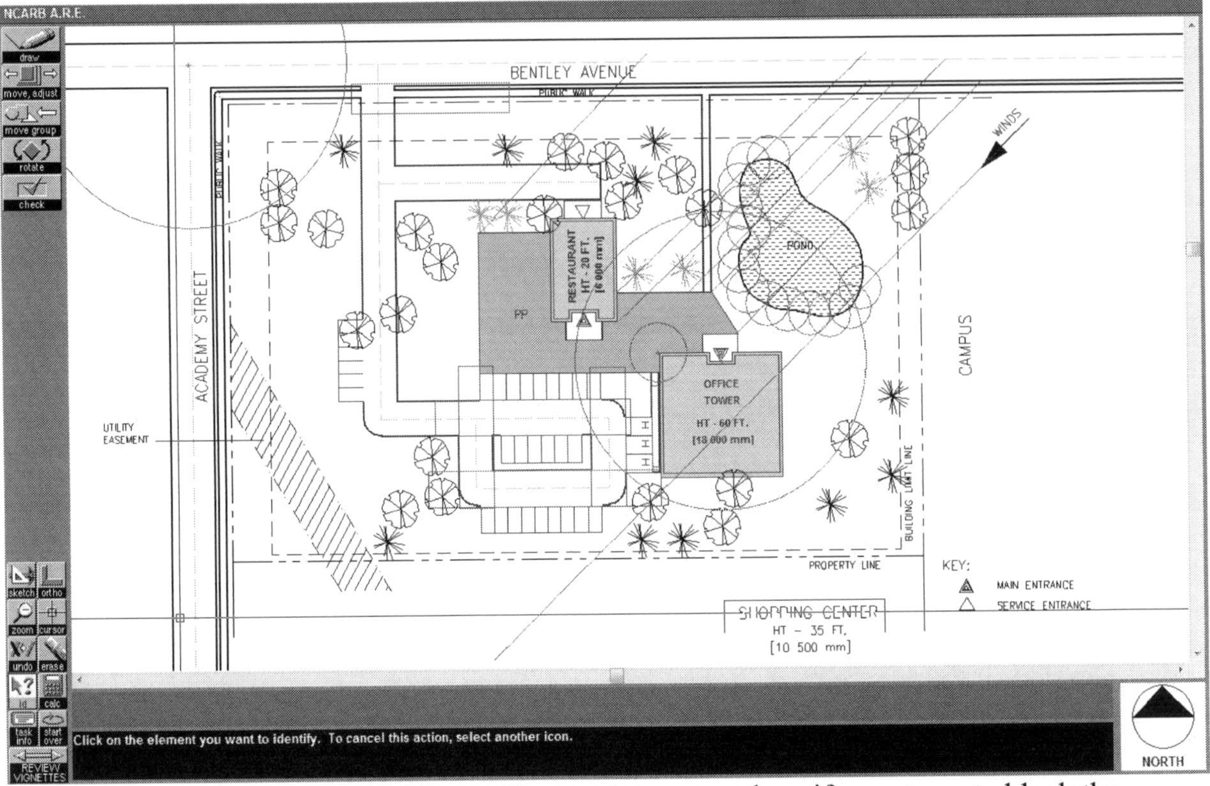

Figure 4.31 Use **Draw > Coniferous Tree** to draw several coniferous trees to block the winds.

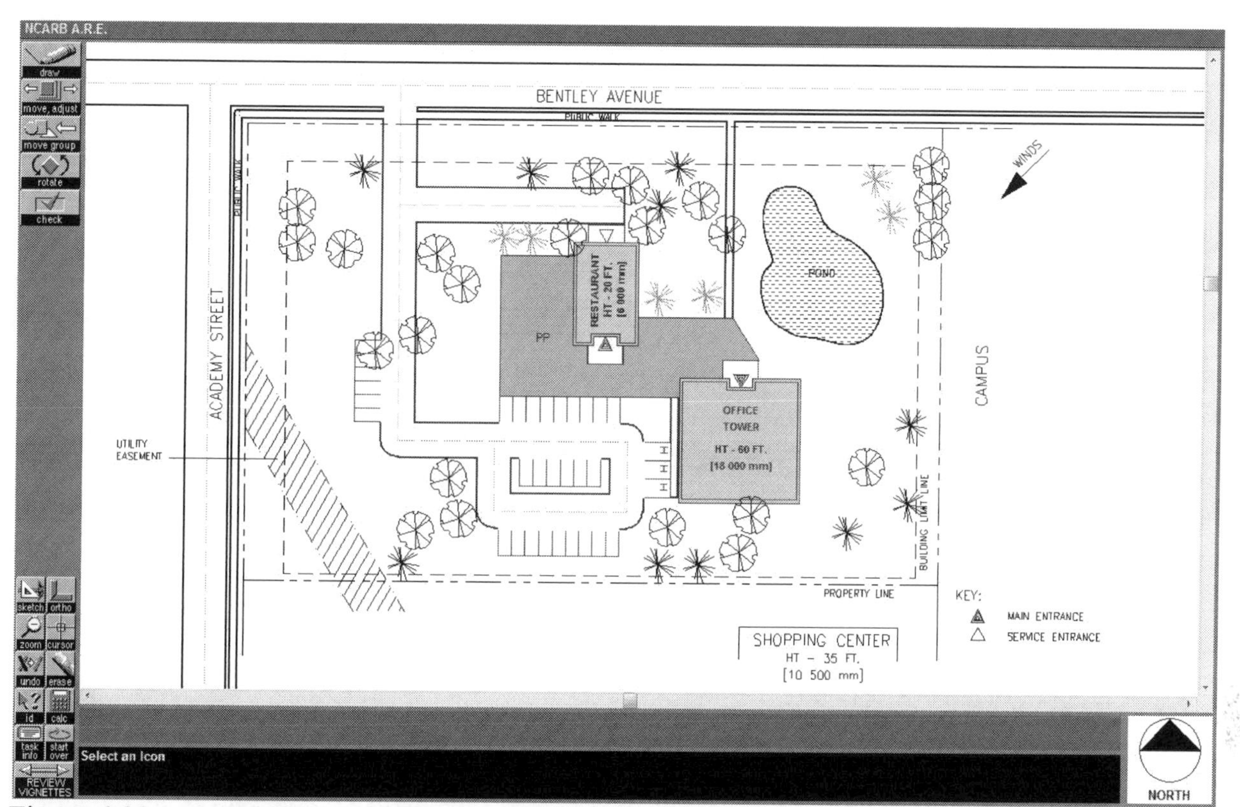

Figure 4.32 Use **Sketch > Hide Sketch Elements** to hide sketch elements.

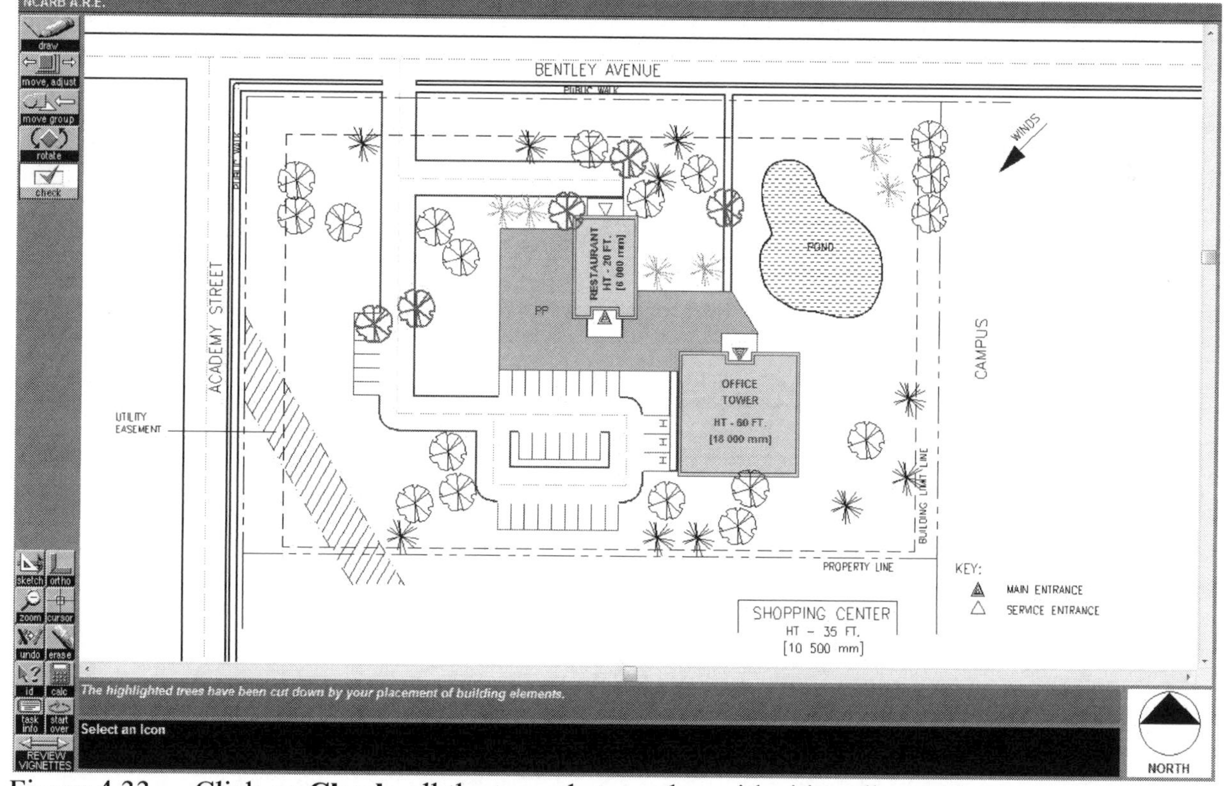

Figure 4.33 Click on **Check**, all the trees that overlap with sidewalk, parking space or other elements will be highlighted.

2. Items that you need to pay special attention to

Several items that you need to pay special attention to:

1) Use **Bubble Diagram** to save time.
2) Use **Sketch** to assist you to locate the elements.
3) A circle with a **15**-ft radius has a 30 ft diameter, which equals the required 30 ft setback from the pond.
4) Use *only* **Draw > Handicap Spaces** to draw handicap spaces. If you accidentally use **Draw >Standard Spaces instead**, you need to erase them and redraw them. There is no way to adjust Standard Spaces and change them into handicap spaces.
5) The intersection of the access drive with the street must be perpendicular to the street for at least the first 20 ft of the drive.
6) The key to draw, align or join different sections of the driveway is to align the centerlines of the driveways.
7) Use **Move, Adjust** to adjust the length of the driveways if necessary.
8) Sometimes it may be easier to erase a driveway, and then redraw it instead of trying to adjust it.
9) The shortest road segment you can draw is 20'-0".
10) Only coniferous trees can block the view all year long. Only coniferous trees can block the prevailing winter winds.
11) Drive-through circulation is required. Dead-end parking is prohibited. Parking along the service drive is prohibited.
12) No construction or built improvements except for driveways and walkways shall occur *outside* the building limit line.
13) Never place buildings *inside* the easement.
14) Pay attention to the size of the Pedestrian Plaza. Draw the Pedestrian Plaza as close to the required size as possible. In any case, it should NOT be 10% larger or smaller than the required size.
15) The sketch lines parallel to the direction of wind are very valuable tool for you to layout the coniferous trees.

Appendixes

A. List of Figures

B. Official Reference Materials Suggested by NCARB

1. General NCARB reference materials for ARE:

Per NCARB, all candidates should become familiar with the latest version of the following codes:

International Code Council, Inc. (ICC, 2006)
International Building Code
International Mechanical Code
International Plumbing Code

National Fire Protection Association (NFPA)
Life Safety Code (NFPA 101)
National Electrical Code (NFPA 70)

National Research Council of Canada
National Building Code of Canada
National Plumbing Code of Canada
National Fire Code of Canada

American Institute of Architects
AIA Documents - 2007

Candidates should be familiar with the Standard on Accessible and Usable Buildings and Facilities (ICC/ANSI A117.1-98)

2. Official NCARB reference materials for the Site Planning & Design (SPD) division:

The Architect's Handbook of Professional Practice
Joseph A. Demkin, AIA, Executive Editor
The American Institute of Architects
John Wiley & Sons, latest edition
A comprehensive book covers all aspect of architectural practice, including two CDs containing the sample AIA contract documents.

Architectural Graphic Standards
Charles G. Ramsey and Harold R. Sleeper
The American Institute of Architects
John Wiley & Sons, latest edition

Canadian Handbook of Practice for Architects,
Committee of Canadian Architectural Councils and The Royal Architectural Institute of Canada, latest edition

Design with Climate
Victor Olgyay
Van Nostrand Reinhold, 1992

Design with Nature
Ian L. McHarg
John Wiley & Sons, 1992

Designing Places for People
C. M. Deasy, FAIA
Whitney Library of Design, 1990

A History of Architecture: Settings & Rituals
Spiro Kostoff
Oxford University Press, 1995

The Image of the City
Kevin Lynch
MIT Press, 1960

Modern Architecture
Alan Colquhoun
Oxford University Press, 2002

The New Urbanism
Peter Katz
McGraw-Hill, 1994

A Pattern Language: Towns, Buildings, Construction
Christopher Alexander, Sarah Ishikawa, and Murray Silverstein
Oxford University Press, 1977

Programming for Design: From Theory to Practice
Edith Cherry
John Wiley & Sons, 1998

Sir Banister Fletcher's A History of Architecture
John Musgrove, Editor
Butterworths-Heinmann, 1996

Site Planning, Third Edition
Kevin Lynch and Gary Hack
MIT Press, 1984

Suburban Nation: The Rise of Sprawl and the

Decline of the American Dream
Andres Duany, Elizabeth Plater-Zybeck, and Jeff Speck
North Point Press, 2001

Sustainable Design Fundamentals for Buildings
National Practice Program
Canada, 2001

C. Other Reference Materials

Chen, Gang. *LEED Green Associate Exam Guide: A Must-Have for the LEED Green Associate Exam: Comprehensive Study Materials, Sample Questions, Mock Exam, Green Building LEED Certification, and Sustainability*. ArchiteG, Inc., latest edition.
This book is a good introduction to green buildings and the LEED building rating systems.

Ching, Francis. *Architecture: Form, Space, & Order.* Wiley, latest edition.
This is one of the best architectural books that you can have. I still flip through it every now and then. It is a great source for inspiration.

Frampton, Kenneth. *Modern Architecture: A Critical History*. Thames and Hudson, London, latest edition.
The publication is a valuable resource for architectural history.

Jarzombek, Mark M. (Author), VikramadityaPrakash (Author), Francis D. K. Ching (Editor). *A Global History of Architecture*. Wiley, latest edition.
Filled with 1000 b & w photos, 50 color photos, and 1500 b & w illustrations, this is a valuable and comprehensive resource for architectural history. It does not limit the topic to a Western perspective, but rather gives a global perspective.

Trachtenberg, Marvin and Isabelle Hyman. *Architecture: From Pre-history to Post-Modernism*. Prentice Hall, Englewood Cliffs, NJ latest edition.
This is also a valuable and comprehensive resource for architectural history.

D. Definition of "Architect" and Some Important Information about Architects and the Profession of Architecture

Architects, Except Landscape and Naval

Significant Points

- About one in five architects are self-employed—more than two times the proportion for all occupations.
- Licensing requirements include a professional degree in architecture, at least three years of practical work training, and passing all divisions of the Architect Registration Examination.
- Architecture graduates may face competition, especially for jobs in the most prestigious firms.

Nature of the Work

People need places in which to live, work, play, learn, worship, meet, govern, shop, and eat. These places may be private or public; indoors or out; rooms, buildings, or complexes, and architects are the individuals who design them. Architects are licensed professionals trained in the art and science of building design who develop the concepts for structures and turn those concepts into images and plans.

Architects create the overall aesthetic and look of buildings and other structures, but the design of a building involves far more than its appearance. Buildings must also be functional, safe, economical, and must suit the needs of the people who use them. Architects consider all these factors when they design buildings and other structures.
Architects may be involved in all phases of a construction project, from the initial discussion with the client through the entire construction process. Their duties require specific skills—designing, engineering, managing, supervising, and communicating with clients and builders. Architects spend a great deal of time explaining their ideas to clients, construction contractors, and others. Successful architects must be able to communicate their unique vision persuasively.

The architect and client discuss the objectives, requirements, and budget of a project. In some cases, architects provide various pre-design services: conducting feasibility and environmental impact studies, selecting a site, preparing cost analysis and land-use studies, or specifying the requirements the design must meet. For example, they may determine space requirements by researching the numbers and types of potential users of a building. The architect then prepares drawings and a report presenting ideas for the client to review.

After discussing and agreeing on the initial proposal, architects develop final construction plans that show the building's appearance and details for its construction. Accompanying these plans are drawings of the structural system; air-conditioning, heating, and ventilating systems; electrical systems; communications systems; plumbing; and, possibly, site and landscape plans. The plans also specify the building materials and, in some cases, the interior furnishings. In developing designs, architects follow building codes, zoning laws, fire regulations, and other ordinances, such as those requiring easy access by people who are disabled. Computer-aided design and drafting (CADD) and Building Information Modeling (BIM) technology has replaced traditional paper and pencil as the most common method for creating design and construction drawings. Continual revision of plans on the basis of client needs and budget constraints is often necessary.

Architects may also assist clients in obtaining construction bids, selecting contractors, and negotiating construction contracts. As construction proceeds, they may visit building sites to make sure that contractors follow the design, adhere to the schedule, use the specified materials, and meet work quality standards. The job is not complete until all construction is finished, required tests are conducted, and construction costs are paid. Sometimes, architects also provide post-construction services, such as facilities management. They advise on energy efficiency measures, evaluate how well the building design adapts to the needs of occupants, and make necessary improvements.
Often working with engineers, urban planners, interior designers, landscape architects, and other professionals, architects in fact spend a great deal of their time coordinating information from, and the work of, other professionals engaged in the same project.
They design a wide variety of buildings, such as office and apartment buildings, schools, churches, factories, hospitals, houses, and airport terminals. They also design complexes such as urban centers, college campuses, industrial parks, and entire communities.
Architects sometimes specialize in one phase of work. Some specialize in the design of one type of building—for example, hospitals, schools, or housing. Others focus on planning and pre-design services or construction management and do minimal design work.

Work environment. Usually working in a comfortable environment, architects spend most of their time in offices consulting with clients, developing reports and drawings, and working with other architects and engineers. However, they often visit construction sites to review the progress of projects. Although most architects work approximately 40 hours per week, they often have to work nights and weekends to meet deadlines.

Training, Other Qualifications, and Advancement
There are three main steps in becoming an architect. First is the attainment of a professional degree in architecture. Second is work experience through an internship, and third is licensure through the passing of the Architect Registration Exam.

Education and training. In most states, the professional degree in architecture must be from one of the 114 schools of architecture that have degree programs accredited by the National Architectural Accrediting Board. However, state architectural registration boards set their own standards, so graduation from a non-accredited program may meet the educational requirement for licensing in a few states.

Three types of professional degrees in architecture are available: a five year bachelor's degree, which is most common and is intended for students with no previous architectural training; a two year master's degree for students with an undergraduate degree in architecture or a related area; and a three or four year master's degree for students with a degree in another discipline.

The choice of degree depends on preference and educational background. Prospective architecture students should consider the options before committing to a program. For example, although the five year bachelor of architecture offers the fastest route to the professional degree, courses are specialized, and if the student does not complete the program, transferring to a program in another discipline may be difficult. A typical program includes courses in architectural history and theory, building design with an emphasis on CADD, structures, technology, construction methods, professional practice, math, physical sciences, and liberal arts. Central to most architectural programs is the design studio, where students apply the skills and concepts learned in the classroom, creating drawings and three-dimensional models of their designs.

Many schools of architecture also offer post-professional degrees for those who already have a bachelor's or master's degree in architecture or other areas. Although graduate education beyond the professional degree is not required for practicing architects, it may be required for research, teaching, and certain specialties.

All state architectural registration boards require architecture graduates to complete a training period—usually at least three years—before they may sit for the licensing exam. Every state, with the exception of Arizona, has adopted the training standards established by the Intern Development Program, a branch of the American Institute of Architects and the National Council of Architectural Registration Boards (NCARB). These standards stipulate broad training under the supervision of a licensed architect. Most new graduates complete their training period by working as interns in architectural firms. Some States allow a portion of the training to occur in the offices of related professionals, such as engineers or general contractors. Architecture students who complete internships while still in school can count some of that time toward the three year training period.

Interns in architectural firms may assist in the design of one part of a project, help prepare architectural documents or drawings, build models, or prepare construction drawings on CADD. Interns also may research building codes and materials or write specifications for building materials, installation criteria, the quality of finishes, and other related details.

Licensure. All states and the District of Columbia require individuals to be licensed (registered) before they may call themselves architects and contract to provide architectural services. During the time between graduation and becoming licensed, architecture school graduates generally work in the field under the supervision of a licensed architect who takes legal responsibility for all work. Licensing requirements include a professional degree in architecture, a period of practical training or internship, and a passing score on all divisions of the Architect Registration Examination. The examination is broken into seven divisions consisting of either multiple choice and/or graphic vignettes. The eligibility period for completion of all divisions is five years from the date of passing your first exam.

Most states also require some form of continuing education to maintain a license, and many others are expected to adopt mandatory continuing education. Requirements vary by state but usually involve the completion of a certain number of credits annually or biennially through workshops, formal university classes, conferences, self-study courses, or other sources.

Other qualifications. Architects must be able to communicate their ideas visually to their clients. Artistic and drawing ability is helpful, but not essential, to such communication. More important are a visual orientation and the ability to understand spatial relationships. Other important qualities for anyone interested in becoming an architect are creativity and the ability to work independently and as part of a team. Computer skills are also required for writing specifications, for two and three dimensional drafting using CADD programs, and for financial management.

Certification and advancement. A growing number of architects voluntarily seek certification by the National Council of Architectural Registration Boards. Certification is awarded after independent verification of the candidate's educational transcripts, employment record, and professional references. Certification can make it easier to become licensed across states. In fact, it is the primary requirement for reciprocity of licensing among state boards that are NCARB members. In 2007, approximately one-third of all licensed architects had this certification.

After becoming licensed and gaining experience, architects take on increasingly responsible duties, eventually managing entire projects. In large firms, architects may advance to supervisory or managerial positions. Some architects become partners in established firms, while others set up their own practices. Some graduates with degrees in architecture also enter related fields, such as graphic, interior, or industrial design; urban planning; real estate development; civil engineering; and construction management.

Employment

Architects held about 132,000 jobs in 2006. Approximately seven out of ten jobs were in the architectural, engineering, and related services industry—mostly in architectural firms with fewer than five workers. A small number worked for residential and nonresidential building construction firms and for government agencies responsible for housing, community planning, or construction of government buildings, such as the U.S. Departments of Defense and Interior, and the General Services Administration. About one in five architects are self-employed.

Job Outlook

Employment of architects is expected to grow faster than the average for all occupations through 2016. Keen competition is expected for positions at the most prestigious firms, and opportunities will be best for those architects who are able to distinguish themselves with their creativity.

Employment change. Employment of architects is expected to grow by 18 percent between 2006 and 2016, which is faster than the average for all occupations. Employment of architects is strongly tied to the activity of the construction industry. Strong growth is expected to come from nonresidential construction as demand for commercial space increases. Residential construction, buoyed by low interest rates, is also expected to grow as more people become homeowners. If interest rates rise significantly, home building may fall off, but residential construction makes up only a small part of architects' work.

Current demographic trends also support an increase in demand for architects. As the population of Sunbelt States continues to grow, the people living there will need new places to live and work. As the population continues to live longer and baby-boomers begin to retire, there will be a need for more healthcare facilities, nursing homes, and retirement communities. In education, buildings at all levels are getting older and class sizes are getting larger. This will require many school districts and universities to build new facilities and renovate existing ones.

In recent years, some architecture firms have outsourced the drafting of construction documents and basic design for large-scale commercial and residential projects to architecture firms overseas. This trend is expected to continue and may have a negative impact on employment growth for lower level architects and interns who would normally gain experience by producing these drawings.

Job prospects. Besides employment growth, additional job openings will arise from the need to replace the many architects who are nearing retirement, and others who transfer to other occupations or stop working for other reasons. Internship opportunities for new architectural students are expected to be good over the next decade, but more students are graduating with architectural degrees and some competition for entry-level jobs can be anticipated. Competition will be especially keen for jobs at the most prestigious architectural firms as prospective architects try to build their reputation. Prospective architects who have had internships while in school will have an advantage in obtaining intern positions after graduation. Opportunities will be best for those architects that are able to distinguish themselves from others with their creativity.

Prospects will also be favorable for architects with knowledge of "green" design. Green design, also known as sustainable design, emphasizes energy efficiency, renewable resources such as energy and water, waste reduction, and environmentally friendly design, specifications, and materials. Rising energy costs and increased concern about the environment has led to many new buildings being built green.

Some types of construction are sensitive to cyclical changes in the economy. Architects seeking design projects for office and retail construction will face especially strong competition for jobs or clients during recessions, and layoffs may ensue in less successful firms. Those involved in the design of institutional buildings, such as schools, hospitals, nursing homes, and correctional facilities, will be less affected by fluctuations in the economy. Residential construction makes up a small portion of work for architects, so major changes in the housing market would not be as significant as fluctuations in the nonresidential market.

Despite good overall job opportunities, some architects may not fare as well as others. The profession is geographically sensitive, and some parts of the Nation may have fewer new building projects. Also, many firms specialize in specific buildings, such as hospitals or office towers, and demand for these buildings may vary by region. Architects may find it increasingly necessary to gain reciprocity in order to compete for the best jobs and projects in other states.

Projections Data

Projections data from the National Employment Matrix

Occupational title	SOC Code	Employment, 2006	Projected employment, 2016	Change, 2006-16		Detailed statistics	
				Number	Percent		
Architects, except landscape and naval	17-1011	132,000	155,000	23,000	18	PDF	zipped XLS

NOTE: Data in this table are rounded. See the discussion of the employment projections table in the *Handbook* introductory chapter on Occupational Information Included in the *Handbook*.

Earnings

Median annual earnings of wage-and-salary architects were $64,150 in May 2006. The middle 50 percent earned between $49,780 and $83,450. The lowest 10 percent earned less than $39,420, and the highest 10 percent earned more than $104,970. Those just starting their internships can expect to earn considerably less.

Earnings of partners in established architectural firms may fluctuate because of changing business conditions. Some architects may have difficulty establishing their own practices and may go through a period when their expenses are greater than their income, requiring substantial financial resources.

Many firms pay tuition and fees toward continuing education requirements for their employees.

For the latest wage information:

The above wage data is from the Occupational Employment Statistics (OES) survey program, unless otherwise noted. For the latest national, state, and local earnings data, visit the following pages:

Architects, except landscape and naval

Related Occupations
Architects design buildings and related structures. Construction managers, like architects, also plan and coordinate activities concerned with the construction and maintenance of buildings and facilities. Others who engage in similar work are landscape architects, civil engineers, urban and regional planners, and designers, including interior designers, commercial and industrial designers, and graphic designers.

Sources of Additional Information

Disclaimer:
Links to non-BLS Internet sites are provided for your convenience and do not constitute an endorsement.

Information about education and careers in architecture can be obtained from:

- The American Institute of Architects, 1735 New York Ave. NW., Washington, DC 20006. Internet: http://www.aia.org
- Intern Development Program, National Council of Architectural Registration Boards, Suite 1100K, 1801 K St. NW., Washington, D.C. 20006. Internet: http://www.ncarb.org OOH ONET Codes 17-1011.00"

Quoted from: Bureau of Labor Statistics, U.S. Department of Labor, Occupational Outlook Handbook, 2008-09 Edition, Architects, Except Landscape and Naval, on the Internet at **http://www.bls.gov/oco/ocos038.htm** (visited November 30, 2008).
Last Modified Date: December 18, 2007

***Note:** Please check the website above for the latest information.*

E. AIA Compensation Survey

Every three years, AIA publishes a Compensation Survey for various positions at architectural firms across the country. It is a good idea to find out the salary before you make the final decision to become an architect. If you are already an architect, it is also a good idea to determine if you are underpaid or overpaid.

See following link for some sample pages for the 2008 AIA Compensation Survey:

http://www.aia.org/aiaucmp/groups/ek_public/documents/pdf/aiap072881.pdf

F. So ... You would Like to Study Architecture

To study architecture, you need to learn how to draft, how to understand and organize spaces and the interactions between interior and exterior spaces, how to do design, and how to communicate effectively. You also need to understand the history of architecture.

As an architect, a leader for a team of various design professionals, you not only need to know architecture, but also need to understand enough of your consultants' work to be able to coordinate with them. Your consultants include soils and civil engineers, landscape architects, structural, electrical, mechanical, and plumbing engineers, interior designers, sign consultants, etc.

There are two major career paths for you in architecture: practice as an architect or teach in college or university. The earlier you determine which path you are going to take, the more likely you will be successful at an early age. Some famous and well-respected architects, like my USC alumnus Frank Gehry, have combined the two paths successfully. They teach at the universities and have their own architectural practice. Even as a college or university professor, people respect you more if you have actual working experience and have some built projects. If you only teach in colleges or universities but have no actual working experience and have no built projects, people will consider you as a "paper" architect, and they are not likely to take you seriously, because they will think you probably do not know how to put a real building together.

In the U.S., if you want to practice architecture, you need to obtain an architect's license. It requires a combination of passing scores on the Architectural Registration Exam (ARE) and eight years of education and/or qualified working experience, including at least one year of working experience in the U.S. Your working experience needs to be under the supervision of a licensed architect to be counted as qualified working experience for your architect's license.

If you work for a landscape architect or civil engineer or structural engineer, some states' architectural licensing boards will count your experience at a discounted rate for the qualification of your architect's license. For example, two years of experience working for a civil engineer may be counted as one year of qualified experience for your architect's license. You need to contact your state's architectural licensing board for specific licensing requirements for your state.

If you want to teach in colleges or universities, you probably want to obtain a master's degree or a Ph.D. It is not very common for people in the architectural field to have a Ph.D. One reason is that there are few Ph.D. programs for architecture. Another reason is that architecture is considered a profession and requires a license. Many people think an architect's license is more important than a Ph.D. degree. In many states, you need to have an architect's license to even use the title "architect," or the terms "architectural" or "architecture" to advertise your service. You cannot call yourself an architect if you do not have an architect's license, even if you have a Ph.D. in architecture. Violation of these rules brings punishment.

To become a tenured professor, you need to have a certain number of publications and pass the evaluation for the tenure position. Publications are very important for tenure track positions. Some people say it is "publish or perish" for the tenured track positions in universities and colleges.

The American Institute of Architects (AIA) is the national organization for the architectural profession. Membership is voluntary. There are different levels of AIA membership. Only licensed architects can be (full) AIA members. If you are an architectural student or an intern but not a licensed architect yet, you can join as an associate AIA member. Contact AIA for detailed information.

The National Council of Architectural Registration Boards (NCARB) is a nonprofit federation of architectural licensing boards. It has some very useful programs, such as IDP, to assist you in obtaining your architect's license. Contact NCARB for detailed

Back Page Promotion

You may be interested in some other books written by Gang Chen:

A. **ARE Mock Exam series. See the following link:**
   http://www.GreenExamEducation.com

B. **LEED Exam Guide series. See the following link:**
   http://www.GreenExamEducation.com

C. ***Building Construction:*** *Project Management, Construction Administration, Drawings, Specs, Detailing Tips, Schedules, Checklists, and Secrets Others Don't Tell You (Architectural Practice Simplified, 2nd edition)*
   http://www.ArchiteG.com

D. ***Planting Design Illustrated***
   http://outskirtspress.com/agent.php?key=11011&page=GangChen

ARE Mock Exam Series

Published ARE books:

Construction Documents and Service (CDS)Are Mock Exam (Architect Registration Exam): ARE Overview, Exam Prep Tips, Multiple-Choice Questions and Graphic Vignettes, Solutions and Explanations
ISBN-13: 9781612650005
(Published May 22, 2011)

Building Design and Construction Systems (BDCS) ARE Mock Exam (Architect Registration Exam): ARE Overview, Exam Prep Tips, Multiple-Choice Questions and Graphic Vignettes, Solutions and Explanations
ISBN-13: 9781612650029
(Published July 12, 2011)

Building Systems (BS) ARE Mock Exam (Architect Registration Exam): ARE Overview, Exam Prep Tips, Multiple-Choice Questions and Graphic Vignettes, Solutions and Explanations
ISBN-13: 9781612650036
(Published October 28, 2011)

Schematic Design (SD) ARE Mock Exam (Architect Registration Exam): ARE Overview, Exam Prep Tips, Graphic Vignettes, Solutions and Explanations
ISBN: 9781612650050
(Published November 18, 2011)

Programming, Planning & Practice (PPP) ARE Mock Exam (Architect Registration Exam): ARE Overview, Exam Prep Tips, Multiple-Choice Questions and Graphic Vignettes, Solutions and Explanations
ISBN-13:9781612650067
(Published May 16, 2012)

Upcoming ARE books:

Other books in the ARE Mock Exam Series are being produced. Our goal is to produce one mock exam book PLUS one guidebook for each of the ARE exam division.

See the following link for the latest information:
http://www.GreenExamEducation.com

LEED Exam Guides series*:* Comprehensive Study Materials, Sample Questions, Mock Exam, Building LEED Certification and Going Green

LEED (Leadership in Energy and Environmental Design) is the most important trend of development, and it is revolutionizing the construction industry. It has gained tremendous momentum and has a profound impact on our environment.

From LEED Exam Guides series, you will learn how to

1. Pass the LEED Green Associate Exam and various LEED AP + exams (each book will help you with a specific LEED exam).

2. Register and certify a building for LEED certification.

3. Understand the intent for each LEED prerequisite and credit.

4. Calculate points for a LEED credit.

5. Identify the responsible party for each prerequisite and credit.

6. Earn extra credit (exemplary performance) for LEED.

7. Implement the local codes and building standards for prerequisites and credit.

8. Receive points for categories not yet clearly defined by USGBC.

There is currently NO official book on the LEED Green Associate Exam, and most of the existing books on LEED and LEED AP are too expensive and too complicated to be practical and helpful. The pocket guides in LEED Exam Guides series fill in the blanks, demystify LEED, and uncover the tips, codes, and jargon for LEED as well as the true meaning of "going green." They will set up a solid foundation and fundamental framework of LEED for you. Each book in the LEED Exam Guides series covers every aspect of one or more specific LEED rating system(s) in plain and concise language and makes this information understandable to all people.

These pocket guides are small and easy to carry around. You can read them whenever you have a few extra minutes. They are indispensable books for all people—administrators; developers; contractors; architects; landscape architects; civil, mechanical, electrical, and plumbing engineers; interns; drafters; designers; and other design professionals.

Why is the LEED Exam Guides series needed?

A number of books are available that you can use to prepare for the LEED Exams:

1. *USGBC Reference Guides*. You need to select the correct version of the *Reference Guide* for your exam.

   The *USGBC Reference Guides* are comprehensive, but they give too much information. For example, *The LEED 2009 Reference Guide for Green Building Design and Construction (BD&C)* has about 700 oversized pages. Many of the calculations in the books are too detailed for the exam. They are also expensive (approximately $200 each, so most people may not buy them for their personal use, but instead, will seek to share an office copy).

   It is good to read a reference guide from cover to cover if you have the time. The problem is not too many people have time to read the whole reference guide. Even if you do read the whole guide, you may not remember the important issues to pass the LEED exam. You need to reread the material several times before you can remember much of it.

   Reading the reference guide from cover to cover without a guidebook is a difficult and inefficient way of preparing for the LEED AP Exam, because you do NOT know what USGBC and GBCI are looking for in the exam.

2. The USGBC workshops and related handouts are concise, but they do not cover extra credits (exemplary performance). The workshops are expensive, costing approximately $450 each.

3. Various books published by a third party are available on Amazon, bn.com and books.google.com. However, most of them are not very helpful.

   There are many books on LEED, but not all are useful.

   LEED Exam Guides series will fill in the blanks and become a valuable, reliable source:

   a. They will give you more information for your money. Each of the books in the LEED Exam Guides series has more information than the related USGBC workshops.

   b. They are exam-oriented and more effective than the USGBC reference guides.

   c. They are better than most, if not all, of the other third-party books. They give you comprehensive study materials, sample questions and answers, mock exams and answers, and critical information on building LEED certification and going green. Other third-party books only give you a fraction of the information.

   d. They are comprehensive yet concise. They are small and easy to carry around. You can read them whenever you have a few extra minutes.

   e. They are great timesavers. I have highlighted the important information that you need to understand and MEMORIZE. I also make some acronyms and short sentences to help you easily remember the credit names.

It should take you about 1 or 2 weeks of full-time study to pass each of the LEED exams. I have met people who have spent 40 hours to study and passed the exams.

You can find sample texts and other information on the LEED Exam Guides series in customer discussion sections under each of my book's listing on Amazon, bn.com and books.google.com.

What others are saying about *LEED GA Exam Guide* (Book 2, LEED Exam Guides series):

"Finally! A comprehensive study tool for LEED GA Prep!

"I took the 1-day Green LEED GA course and walked away with a power point binder printed in very small print—which was missing MUCH of the required information (although I didn't know it at the time). I studied my little heart out and took the test, only to fail it by 1 point. Turns out I did NOT study all the material I needed to in order to pass the test. I found this book, read it, marked it up, retook the test, and passed it with a 95%. Look, we all know the LEED GA exam is new and the resources for study are VERY limited. This one's the VERY best out there right now. I highly recommend it."
—ConsultantVA

"Complete overview for the LEED GA exam

"I studied this book for about 3 days and passed the exam … if you are truly interested in learning about the LEED system and green building design, this is a great place to start."
—K.A. Evans

"A Wonderful Guide for the LEED GA Exam

"After deciding to take the LEED Green Associate exam, I started to look for the best possible study materials and resources. From what I thought would be a relatively easy task, it turned into a tedious endeavor. I realized that there are vast amounts of third-party guides and handbooks. Since the official sites offer little to no help, it became clear to me that my best chance to succeed and pass this exam would be to find the most comprehensive study guide that would not only teach me the topics, but would also give me a great background and understanding of what LEED actually is. Once I stumbled upon Mr. Chen's book, all my needs were answered. This is a great study guide that will give the reader the most complete view of the LEED exam and all that it entails.

"The book is written in an easy-to-understand language and brings up great examples, tying the material to the real world. The information is presented in a coherent and logical way, which optimizes the learning process and does not go into details that will not be needed for the LEED Green Associate Exam, as many other guides do. This book stays dead on topic and keeps the reader interested in the material.

"I highly recommend this book to anyone that is considering the LEED Green Associate Exam. I learned a great deal from this guide, and I am feeling very confident about my chances for passing my upcoming exam."
—PavelGeystrin

"Easy to read, easy to understand

"I have read through the book once and found it to be the perfect study guide for me. The author does a great job of helping you get into the right frame of mind for the content of the exam. I had started by studying the Green Building Design and Construction reference guide for LEED projects produced by the USGBC. That was the wrong approach, simply too much information with very little retention. At 636 pages in textbook format, it would have been a daunting task to get through it. Gang Chen breaks down the points, helping to minimize the amount of information but maximizing the content I was able to absorb. I plan on going through the book a few more times, and I now believe I have the right information to pass the LEED Green Associate Exam."
—Brian Hochstein

"All in one—LEED GA prep material

"Since the LEED Green Associate exam is a newer addition by USGBC, there is not much information regarding study material for this exam. When I started looking around for material, I got really confused about what material I should buy. This LEED GA guide by Gang Chen is an answer to all my worries! It is a very precise book with lots of information, like how to approach the exam, what to study and what to skip, links to online material, and tips and tricks for passing the exam. It is like the 'one stop shop' for the LEED Green Associate Exam. I think this book can also be a good reference guide for green building professionals. A must-have!"
—SwatiD

"An ESSENTIAL LEED GA Exam Reference Guide

"This book is an invaluable tool in preparation for the LEED Green Associate (GA) Exam. As a practicing professional in the consulting realm, I found this book to be all-inclusive of the preparatory material needed for sitting the exam. The information provides clarity to the fundamental and advanced concepts of what LEED aims to achieve. A tremendous benefit is the connectivity of the concepts with real-world applications.

"The author, Gang Chen, provides a vast amount of knowledge in a very clear, concise, and logical media. For those that have not picked up a textbook in a while, it is very manageable to extract the needed information from this book. If you are taking the exam, do yourself a favor and purchase a copy of this great guide. Applicable fields: Civil Engineering, Architectural Design, MEP, and General Land Development."
—Edwin L. Tamang

Note: Other books in the **LEED Exam Guides seriesare** in the process of being produced. At least **One book will eventually be produced for each of the LEED exams.** The series include:

LEED GA EXAM GUIDE: *A Must-Have for the LEED Green Associate Exam: Comprehensive Study Materials, Sample Questions, Mock Exam, Green Building LEED Certification, and Sustainability*(3rd Large Format Edition),LEED Exam Guides series, ArchiteG.com (Published January 3, 2011)

LEED GA MOCK EXAMS: *Questions, Answers, and Explanations: A Must-Have for the LEED Green Associate Exam, Green Building LEED Certification, and Sustainability*, LEED Exam Guides series, ArchiteG.com (Published August 6, 2010)

LEED BD&C EXAM GUIDE: *A Must-Have for the LEED AP BD+C Exam: Comprehensive Study Materials, Sample Questions, Mock Exam, Green Building Design and Construction, LEED Certification, and Sustainability* (2nd Edition), LEED Exam Guides series, ArchiteG.com (Published December 26, 2011)

LEED BD&C MOCK EXAMS: *Questions, Answers, and Explanations: A Must-Have for the LEED AP BD+C Exam, Green Building LEED Certification, and Sustainability*, LEED Exam Guides series, ArchiteG.com (Published November 26, 2011)

LEEDAP Exam Guide: *Study Materials, Sample Questions, Mock Exam, Building LEED Certification (LEED NC v2.2), and Going Green,* LEED Exam Guides series, LEEDSeries.com (Published on 9/23/2008).

LEED ID&C EXAM GUIDE: *A Must-Have for the LEED AP ID+C Exam: Comprehensive Study Materials, Sample Questions, Mock Exam, Green Interior Design and Construction, LEED Certification, and Sustainability,* LEED Exam Guides series, ArchiteG.com (Published March 8, 2010)

LEED O&M MOCK EXAMS: *Questions, Answers, and Explanations: A Must-Have for the LEED O&M Exam, Green Building LEED Certification, and Sustainability*, LEED Exam Guides series, ArchiteG.com (Published September 28, 2010)

LEED O&M EXAM GUIDE: *A Must-Have for the LEED AP O+M Exam: Comprehensive Study Materials, Sample Questions, Mock Exam, Green Building Operations and Maintenance, LEED Certification, and Sustainability(LEED v3.0)*,LEED Exam Guides series, ArchiteG.com

LEEDHOMES EXAM GUIDE: *A Must-Have for the LEED AP Homes Exam: Comprehensive Study Materials, Sample Questions, Mock Exam, Green Building LEED Certification, and Sustainability*, LEED Exam Guides series, ArchiteG.com

LEEDND EXAM GUIDE: *A Must-Have for the LEED AP Neighborhood Development Exam: Comprehensive Study Materials, Sample Questions, Mock Exam, Green Building LEED Certification, and Sustainability*, LEED Exam Guides series, ArchiteG.com

How to order these books:
You can order the books listed above at:
http://www.GreenExamEducation.com

OR
http://www.ArchiteG.com

Building Construction

Project Management, Construction Administration, Drawings, Specs, Detailing Tips, Schedules, Checklists, and Secrets Others Don't Tell You (Architectural Practice Simplified, 2nd edition)

Learn the Tips, Become One of Those Who Know Building Construction and Architectural Practice, and Thrive!

For architectural practice and building design and construction industry, there are two kinds of people: those who know, and those who don't. The tips of building design and construction and project management have been undercover—until now.

Most of the existing books on building construction and architectural practice are too expensive, too complicated, and too long to be practical and helpful. This book simplifies the process to make it easier to understand and uncovers the tips of building design and construction and project management. It sets up a solid foundation and fundamental framework for this field. It covers every aspect of building construction and architectural practice in plain and concise language and introduces it to all people. Through practical case studies, it demonstrates the efficient and proper ways to handle various issues and problems in architectural practice and building design and construction industry.

It is for ordinary people and aspiring young architects as well as seasoned professionals in the construction industry. For ordinary people, it uncovers the tips of building construction; for aspiring architects, it works as a construction industry survival guide and a guidebook to shorten the process in mastering architectural practice and climbing up the professional ladder; for seasoned architects, it has many checklists to refresh their memory. It is an indispensable reference book for ordinary people, architectural students, interns, drafters, designers, seasoned architects, engineers, construction administrators, superintendents, construction managers, contractors, and developers.

You will learn:

1. How to develop your business and work with your client.
2. The entire process of building design and construction, including programming, entitlement, schematic design, design development, construction documents, bidding, and construction administration.
3. How to coordinate with governing agencies, including a county's health department and a city's planning, building, fire, public works departments, etc.
4. How to coordinate with your consultants, including soils, civil, structural, electrical, mechanical, plumbing engineers, landscape architects, etc.
5. How to create and use your own checklists to do quality control of your construction documents.
6. How to use various logs (i.e., RFI log, submittal log, field visit log, etc.) and lists (contact list, document control list, distribution list, etc.) to organize and simplify your work.
7. How to respond to RFI, issue CCDs, review change orders, submittals, etc.
8. How to make your architectural practice a profitable and successful business.

Planting Design Illustrated

A Must-Have for Landscape Architecture: A Holistic Garden Design Guide with Architectural and Horticultural Insight, and Ideas from Famous Gardens in Major Civilizations

One of the most significant books on landscaping!

This is one of the most comprehensive books on planting design. It fills in the blanks of the field and introduces poetry, painting, and symbolism into planting design. It covers in detail the two major systems of planting design: formal planting design and naturalistic planting design. It has numerous line drawings and photos to illustrate the planting design concepts and principles. Through in-depth discussions of historical precedents and practical case studies, it uncovers the fundamental design principles and concepts, as well as the underpinning philosophy for planting design. It is an indispensable reference book for landscape architecture students, designers, architects, urban planners, and ordinary garden lovers.

What Others Are Saying About *Planting Design Illustrated* …

"I found this book to be absolutely fascinating. You will need to concentrate while reading it, but the effort will be well worth your time."
—Bobbie Schwartz, former president of APLD (Association of Professional Landscape Designers) and author of *The Design Puzzle: Putting the Pieces Together*.

"This is a book that you have to read, and it is more than well worth your time. Gang Chen takes you well beyond what you will learn in other books about basic principles like color, texture, and mass."
—Jane Berger, editor & publisher of gardendesignonline

"As a longtime consumer of gardening books, I am impressed with Gang Chen's inclusion of new information on planting design theory for Chinese and Japanese gardens. Many gardening books discuss the beauty of Japanese gardens, and a few discuss the unique charms of Chinese gardens, but this one explains how Japanese and Chinese history, as well as geography and artistic traditions, bear on the development of each country's style. The material on traditional Western garden planting is thorough and inspiring, too. *Planting Design Illustrated* definitely rewards repeated reading and study. Any garden designer will read it with profit."
—Jan Whitner, editor of the *Washington Park Arboretum Bulletin*

"Enhanced with an annotated bibliography and informative appendices, *Planting Design Illustrated* offers an especially "reader friendly" and practical guide that makes it a very strongly recommended addition to personal, professional, academic, and community library gardening & landscaping reference collection and supplemental reading list."
—Midwest Book Review

“Where to start? *Planting Design Illustrated* is, above all, fascinating and refreshing! Not something the lay reader encounters every day, the book presents an unlikely topic in an easily digestible, easy-to-follow way. It is superbly organized with a comprehensive table of contents, bibliography, and appendices. The writing, though expertly informative, maintains its accessibility throughout and is a joy to read. The detailed and beautiful illustrations expanding on the concepts presented were my favorite portion. One of the finest books I’ve encountered in this contest in the past 5 years.”
—Writer’s Digest 16th Annual International Self-Published Book Awards Judge’s Commentary

“The work in my view has incredible application to planting design generally and a system approach to what is a very difficult subject to teach, at least in my experience. Also featured is a very beautiful philosophy of garden design principles bordering poetry. It’s my strong conviction that this work needs to see the light of day by being published for the use of professionals, students & garden enthusiasts.”
—Donald C. Brinkerhoff, FASLA, chairman and CEO of Lifescapes International, Inc.

Index

Made in the USA
Lexington, KY
14 February 2015